New Approaches to Human Reproduction

New Approaches to Human Reproduction

Social and Ethical Dimensions

EDITED BY
Linda M. Whiteford
and Marilyn L. Poland

Westview Press
BOULDER & LONDON

To the pleasure
and perpetuation
of the secret
Mink Society

Chapter 2, by Rayna Rapp, appeared in *Medical Anthropology Quarterly* 2:2 (1988). It is reprinted here, with slight rewording, by permission of the American Anthropological Association.

Published in 1989 in the United States of America by Westview Press, Inc., 5500 Central Avenue, Boulder, Colorado 80301, and in the United Kingdom by Westview Press, Inc., 13 Brunswick Centre, London WC1N 1AF, England

Library of Congress Cataloging-in-Publication Data
New approaches to human reproduction : social and ethical dimensions /
edited by Linda M. Whiteford and Marilyn L. Poland.
p. cm.
Bibliography: p.
Includes index.
ISBN 0-8133-0450-4
1. Pregnancy—Moral and ethical aspects. 2. Human reproduction—Moral and ethical aspects. 3. Infants (Newborn)—Medical care—Moral and ethical aspects. 4. Surrogate mothers. 5. Human reproduction—Technological innovations—Moral and ethical aspects.
I. Whiteford, Linda M. II. Poland, Marilyn L. (Marilyn Laken)
RG560.N48 1989
174′.2—dc19
88-14398
CIP

Printed and bound in the United States of America

The paper used in this publication meets the requirements of the American National Standard for Permanence of Paper for Printed Library Materials Z39.48-1984.

10 9 8 7 6 5 4 3 2 1

Contents

Preface

This book grew out of our concern with what we saw as a major gap between cultural values and medical technology, most dramatically focused in the areas of conception, birth, and neonatality. In order to begin to address that gap, social scientists studying health and reproduction were brought together in November 1985 at the American Anthropological Association meeting in Washington, D.C., to present their research. In March of 1986, a similar group of experts came together in Reno, Nevada, to give their views at the meeting of the Socity for Applied Anthropology. The 1985 and 1986 symposia reflected a small segment of a larger group of social scientists concerned with ethical dilemmas generated by the interface of cultural values, medical technology, and human reproduction. This book brings together current research data and analysis particularly relevant for social scientists as well as nurses, public health professionals, and physicians. By drawing on experts in social science, ethics, medicine, and law, we have tried to provide a multidisciplinary perspective on some of the social and ethical issues generated by the new reproductive technologies.

We want to thank several people who helped put this book together, particularly our editorial assistant, Lois Randolph. Lois carefully read each chapter and gave us the benefit of her knowledge and insight. Diane Rambeau and Patricia Miller-Shaivitz, both graduate students in the Department of Anthropology at the University of South Florida; Rebecca Miller-Jones, graduate secretary in the Department of Anthropology at the University of South Florida; and Pamela Wilson from Wayne State University also labored long and well checking notes, correcting grammar, and finding missing information in the bibliographies. Michael Copeland, Nita Desai, and Peter Feller from the Word Processing Center at the University of South Florida and Pamela Wilson from the Department of Obstetrics and Gynecology at Wayne State University facilitated the completion of this book by putting it all on computer tape. We appreciate their help and the contribution of their time, which was made possible by the generosity of the offices of the Dean of Social Sciences at the University of South Florida and the Chairman of Obstetrics and Gynecology at Wayne State University. We also thank our editor, Dean Birkenkamp, for always being accessible to us and supportive of this project. Although we cannot thank the anonymous reviewer by name, those comments provided excellent suggestions—most of which we took seriously and used. Although we are very pleased with the results of the work of all these people, the faults, oversights, and omissions remain ours.

Linda M. Whiteford
Marilyn L. Poland

Introduction

Linda M. Whiteford and Marilyn L. Poland

Recent changes in reproductive technology have altered control of fertility, childbearing, and child rearing and have changed the social definitions of fetus and parenthood. As a result, serious ethical issues are being raised regarding rights and responsibilities of the parties involved. Although we are better able to identify medical problems, we still do not know their social implications nor can we guarantee their desired outcomes. Often, mechanically generated information places people in impossible positions in which they must decide whether to prolong the life and pain of a catastrophically sick infant, to abort a supposedly defective fetus, to use another person exclusively to create a child, or to choose between the rights of a woman and those of her unborn child. In this book, we look at the social contexts in which these decisions are made in order to identify patterns of pressure and influence that affect the decision-makers. Medical decisions affecting reproductive abilities, neonatal survival, prenatal testing, and genetic counseling will never be easy, but they must be made with an awareness of the social costs. These costs, we will argue, become clearer by understanding the effect of personal, organizational, and cultural influences.

This introduction is divided into two parts: (1) a discussion of the three themes that underlie the book and (2) a description of how the book is organized. We begin by establishing general sequelae of reproductive technology: changes in control over reproduction, social consequences of reproductive technology, and ethical dilemmas that result.

Changes in Control of Reproduction

For thousands of years, human females controlled their fertility primarily through cultural practices (prolonged breastfeeding and sanctions against sexual intercourse following birth) and by supernatural means (invoking the goodwill of powerful deities). The course of pregnancy and birth was under similar control. Pregnancy was generally viewed as a woman-centered condition. The concept of a "fetus" represented a potential versus a real being (that is, spirits invaded the body and lived there during pregnancy). And pregnancy itself could result in a living being whose social designation as "child" could vary in time from birth to one year or more later. Birth was also woman-centered and primarily woman-attended; midwives or experienced female kin generally attended birthing women.

The social definition of pregnancy and birth as a natural, woman-centered condition has undergone a dramatic change over the past four centuries. Around 1550, male physicians began to compete with midwives for control over the birth process. Birth technology, such as the use of forceps, was introduced by the medical practitioners of that time. Birth practices based on thousands of years of woman-experience, such as squatting or sitting to give birth have given way to physician-preferred procedures, such as use of the lithotomy position (lying on one's back with legs elevated). King Louis XIV initiated the use of this biologically unnatural pose by women in childbirth so that he could have a better view of one of his favorite events. Pregnancy itself is no longer defined as a normal, natural condition but as an illness requiring medical management.

Women sought physicians and technology rather than midwives because of their growing belief that medical control was superior to control through natural or supernatural means. And, indeed, introduction of modern obstetrics has contributed to safer, healthier birth experiences for women and their infants. In addition, the development of modern methods of contraception and abortion, along with infertility diagnosis and treatment, have given women a form of control over their own fertility that they never had before. Improvements in the ability to diagnose genetic conditions and in the medical care of sick newborns add to the major changes introduced by medical technology.

Although technology has offered women more control over the timing of their pregnancies, has made the process of reproduction safer, and has saved many sick newborns, there have been social costs as well. These costs include women's simultaneous loss of control over some decisions about pregnancy and the care of ill newborns, increased risk of injury because of side effects of medical or surgical procedures, and the potential for inaccurate information leading to unnecessary and even harmful medical interventions.

Women's simultaneous gain and loss of control over reproduction is seen most clearly in the shift in medical and social attention away from the woman to her fetus/child. The shift began as maternal mortality rates fell more rapidly than infant mortality rates, thus focusing attention on the more vulnerable fetus/newborn. The shift in attention was accelerated when ultrasonography, the use of sound waves to produce a computer-enhanced picture of the fetus, altered the social definition of fetus from "potential" to "real," with concomitant legal arguments for fetal rights. This has placed the rights of the mother in conflict with those of her fetus when the health or life of one is jeopardized.

Women's control over reproduction and parenthood has also been challenged by considerations of social status, particularly when status is linked with conception technology, such as artificial insemination as used in surrogacy arrangements. In the "Baby M" case, for example, the judge upheld the contract between the biological mother and the more affluent biological father and his wife (the social mother). This legal decision gave control over the child to the biological father, who initially purchased the contract. Thus, status differences, defined by wealth and education, also play a role in who gets pregnant and who controls pregnancy and child rearing.

Women and infants have largely benefited from modern medicine, but the use of medical technology is also associated with risks of harmful effects. For example, although more sick newborns are being saved, some of these infants will bear the scars of the technology that allowed them to live. Cognitive and emotional deficits are often associated with some severely ill infants. Chronic medical problems, some requiring the need for expensive and difficult treatments, such as respirator assistance at home, may also result. The growing number of these children in the United States reflects the increased use of technology and a change in social and medical attitudes toward the sick newborn from one of little interference with the natural course to doing everything possible to save the child's life. This change in attitude has had an impact on parental decisionmaking in two ways. First, decisions by parents to withhold potentially lifesaving technology are being overridden by the courts, despite the fact that the parents are the ones responsible for the continuing care of the child with a chronic condition. Second, there is great potential to communicate bias in favor of intervention when parents do have a choice, such as in genetic counseling. In both instances, parental participation in decisions about an abnormal fetus or newborn is encouraged, if not required; yet the power of technology is so compelling that decisions based on it may, at times, override parental preferences.

The final cost of technology is the potential for false and inaccurate information leading to decisions that are not in the best interests of mother or infant. This is best seen in several cases of court-ordered cesarean section in which placenta previa (location of the placenta over the cervical os, or opening to the uterus) was misdiagnosed and the mother ordered to undergo surgery to save the life of her fetus, only to have a normal vaginal birth subsequently. Thus, increasingly, technology and those who control it have the final word in determining the "nature" of pregnancy, birth, and the care of sick newborns.

Social Consequences of Reproductive Technology

The new reproductive technologies, such as in vitro fertilization, artificial insemination by donor (AID), surrogacy, and embryo transfer have followed the introduction of dependable birth-control methods. Sex without reproduction was made possible by reliable contraception. Now, medicine and commerce offer reproduction without sex. Completing the separation between sex and reproduction, and between conception and sexual intercourse, opens new dimensions of legal, ethical, and social consequences.

Roughly one in every six married couples of childbearing age in the United States is infertile. The availability of technologically assisted conception and the increasing number of couples affected by infertility suggests that recourse to medical interventions will increase. More couples will enter into agreements with lawyers, psychiatrists, and others to broker their fertility in order to provide them with a child.

Knowledge of fertility brokering has resulted in social controversy by raising public questions about private topics. Public opinion, reflecting changing private views, has come to accept some forms of reproductive technology while still debating other forms. In 1973, when John and Doris Zio attempted to become the first parents of an in vitro fertilized baby, the attending physician was rebuked by his superior for engaging in what was then considered "dangerous and unethical practices," and the incubator was opened, killing the fertilized embryo. Practices that were unacceptable to the supervising physician and Columbia-Presbyterian Hospital in New York in 1973 were no longer considered unethical ten years later in 1983, when the same supervising physician was made the director of the hospital's new in vitro fertilization program. Social opinion had changed sufficiently so as to reclassify the concept of in vitro fertilization; what had been considered a "dangerous and unethical" process had become medically and socially acceptable.

Changing public opinion and increasingly sophisticated medical technology raise questions about reproduction never before asked. Surrogacy, for example, forces society to consider its definitions of motherhood. Artificial insemination by donor focuses attention on questions of fatherhood. Rights of physicians, spouses, and even frozen embryos now have to be decided.

In 1979, Illinois became the first state to deal explicitly with legal issues surrounding in vitro fertilization, deciding to make any physician involved in in vitro fertilization the legal guardian of the resultant embryo and liable for prosecution under an 1877 law prohibiting child abuse. In France, the question of who owned donated sperm became a legal issue when Corrine

Parpalaix's husband died. Mr. Parpalaix had earlier donated sperm; after his death his widow requested some of the sperm so that she could become pregnant. The sperm bank refused. The man had left no instructions for his sperm, they said. The widow sued the sperm bank for possession of the dead man's seed. The state prosecutor, siding with the sperm bank, argued that the wife had no more right to the dead man's sperm than she had to his feet or ears. The court, to the surprise of the legal experts, decided in favor of the wife, giving Parpalaix her husband's deposit.

As this case shows, the new reproductive technology makes possible unusual questions of ownership and obligations. How should sperm ownership be defined? Is it similar to an organ transplant? Should sperm be thought of as inheritable property? Who should have access to someone's sperm, and who should have the responsibility for its physical protection and the well-being of the resultant child? These are social issues generated by culturally engendered needs and assisted by technological possibilities. Too often, there are no guidelines in place to help with the determination of social decisions. And traditional responses no longer fit the questions.

Who, for instance, will decide the fate of viable embryos orphaned before they are born? When both Mario and Elsa Rios were killed in a plane crash in 1981, they left behind fertilized embryos being stored for future implantation. The embryos are currently being stored in a freezer in Melbourne, Australia. The Australian government has given permission for them to be destroyed but is reluctant to act on that permission. Their destruction would be the final termination of the Rios's reproductive potential. Under ordinary circumstances, the Rios's ability to reproduce would have unavoidably ended with their deaths. No further human decisions would have been required. Now, however, humans must decide what should be done with the Rios's embryos. Should they be destroyed, saved forever, given to some family member? These are some of the extraordinary questions forced on ordinary people by the new reproductive technologies.

Ethical Dilemmas Incurred by Technology

Recent developments in reproductive technology—such as genetic testing, infertility diagnosis and tratment, maternal and fetal testing, and the care of sick newborns—challenge our current sets of values and principles relating to fertility, maternal responsibility, and parenthood. New technology is changing policy and practice in reproductive and perinatal health care, which in turn change perceptions of patients' choices, autonomy, and physician obligation. Such fundamental changes in this aspect of our culture, and the social implications of those changes, need to be further investigated and better understood.

Many of the dilemmas faced by men and women, as either parents or medical practitioners, stem from the fact that there are no appropriate guidelines for the decisions they must now make. Nor are there guidelines for who should make the decisions. For example, should anyone have the right and responsibility to force an unwilling woman to undergo major surgery (cesarean section), a decision frequently based on technologically generated, but often ambiguous, data? How should decisions be made about the extent of neonatal surgery and the extended use of life supports for critically ill newborns, and who should make those decisions? Whose rights should be most respected in commercial surrogacy, those of the surrogate mother, the biological father, or the child? How should those rights be defined: by legal standards, social mores, or abstract ethical principles? In order to address these dilemmas, we need a clearer understanding of the social implications of technologically assisted reproduction.

The contributors to this book discuss a wide variety of issues surrounding reproduction, including the private and personal domain of infertility, the inability to obtain prenatal care or to choose a birth method, and decisions about how to treat dying infants. In each of these areas, changes in technology have propelled people into unfamiliar situations in which their actions have major and irreversible consequences.

Organization

Intended as an introduction to social and ethical issues surrounding technologically assisted reproduction from an interdisciplinary perspective, this book is organized around the processes and outcomes of decisions surrounding pregnancy and birth. Within these topics are themes of informed consent, patient autonomy, and social inequities in the distribution of power in crisis decisionmaking. It is not meant to be a comprehensive examination of all of the issues but, rather, focuses on several critical topics. Each of the three sections addresses ethical dilemmas created by a particular set of conflicts between technology and human limitations: (1) problems of quality and access to health care during pregnancy and birth, (2) changes in decision-making in treating severely ill newborns, and (3) consequences of family formation by artificial insemination—in particular, the case of commercial surrogacy.

Part One, "Ethics of Quality, Access, and Care During Pregnancy," deals with technology-engendered ethical quandaries in the delivery of quality care before and during birth. The section opens with an analysis by Brigitte Jordan and Susan Irwin of case histories of women who were forced to undergo unwanted cesarean sections. This critical review of the medical and legal decisions made on behalf of the unborn fetus over the objections of the mother reinforces the need to further examine the exercise of power and control over maternal decisionmaking. Rayna Rapp discusses social and cultural barriers to communication during genetic counseling and raises questions about possible bias introduced by the counselor. Marilyn Poland completes this section with her data on sociocultural barriers to prenatal care experienced by indigent women. Two ethical questions are explored: social obligations to provide prenatal care and rights and responsibilities of pregnant women in regard to seeking prenatal care.

Part Two, "Ethical Decisions in the Treatment of Newborns," explores the difficult topic of decision making in the care of catastrophically ill neonates. Renee R. Anspach describes how families and providers are caught in a web of social expectations based on class, gender, and role stereotypes. The neonatal intensive-care unit (NICU) is a social, as well as medical, unit that needs to be viewed as having its own organizational hierarchy. Anspach contends that decisions that affect infants in that unit cannot fully be understood without cognizance of the unit's particular social organization.

Jeanne Guillemin et al. and Betty Levin take their views of decision making in the NICU from the perspective of the medical practitioner, as they document factors that influence changes in physician's attitudes concerning treatment of severely ill newborns. As Anspach details how social organization affects decision making, Levin and Guillemin et al. record how public pressures and resultant policy changes influence decision making. Levin's consideration of the "Baby Doe" case makes that effect clear.

Part Three, "Ethical Implications of Family Formation by Surrogacy," investigates the social consequences of technologically assisted fertilization, particularly in the case of commercial surrogacy. Kamran Moghissi presents a thorough description of the technology used in artificial insemination. Judith Lasker and Susan Borg tie social response, such as secrecy, to the type of artificial insemination chosen. They report that couples who select artificial insemination by donor are counseled to keep the insemination a secret and often do, thereby placing a barrier of secrecy between parents and children. Couples whose babies are born through a surrogate mother choose more often to acknowledge that process than do parents in non-surrogate AID births.

Linda Whiteford examines social issues underlying commercial surrogacy, emphasizing both the real and potential consequences of this procedure on the family as a social unit. She raises serious concerns about its impact on the surrogate mother, contracting couple, and offspring. Sandra Garcia provides an analysis of the legal entanglements of commercial surrogacy. In applying the Sales Law Act to the surrogacy process, Garcia points out the moral and ethical difficulties of legal protection. "Baby Cotton," the first baby born through commercial surrogacy in the United Kingdom, and the effect of its birth in changing public opinion are described by Sybil Wolfram. Wolfram examines objections to commercial surrogacy raised in England.

In a second chapter by Garcia, Baby M is presented as a case of undefined rights, unenforceable responsibilities, and inadequate remedies. Garcia traces the development and conclusion of the New Jersey trial concerning Baby M and applies her analysis of the Sales Law Act to the case.

Underlying Themes

Informed consent, "that particularly American legal-medical document" according to Rapp, may be less a process of informing participants of the range of medical options available to them in which they are principal decision-makers than it is the process of securing the participants' agreement with decisions already made by the medical staff. But in theory, informed consent implies active participation of the patient, and/or the parents, in decision making about options in medical treatment; that decision, based on informed knowledge, also implies the choice, as Thomas Szasz has suggested, to choose *not* to be a patient. That women may be forced to have cesarean sections by court order forces us to confront ethical quandaries generated by conflicting views of patient rights and to raise questions about the meaning of informed consent. Less dramatic but no less personally devastating is the difficulty in securing informed consent because of cultural and communicative problems between genetic counselors and their clients.

Patient, or client, autonomy is akin to and, in fact, embedded in the concept of informed consent. Autonomy, along with nonmalfeasance, beneficence, and justice form a cluster of basic ethical principles that are commonly applied to bioethical analyses. The concept of patient autonomy often comes into sharp conflict with the practice of physician authority; this conflict is exacerbated when the physician trusts technologically produced information more than the patient does. Difficulties concerning patient autonomy are further complicated by public readiness to seek redress in the courts for both avoidable and unavoidable injuries.

Socioeconomic, cultural, and gender differences direct *the distribution of power in medical decision making* and access to medical services. Free or low-cost prenatal services may be available to poor women, but there are class-related barriers to their use. Although obvious problems, such as difficulties in scheduling and transportation, may be overcome through informed planning, the roots of intimidation by the medical process, the inability to read the required forms, and a general lack of motivation are deeply embedded in the structure of social inequality. Women who have not accepted the Western medical model may be isolated from financial resources and from social and informational support and, as a consequence, may not be considered "good" patients and may not be approved of by the medical staff. A conflict exists between those who have a social (or "folk") definition of an event and those who have a medical conceptualization;

this conflict often falls along lines that divide groups by class, race, gender, and culture.

Power is also unequally distributed between members of a procreative couple, between providers and clients, between rich and poor, and between physicians and nurses. Improved medical procedures in neonatal intensive care have changed physicians' attitudes in favor of increased medical intervention, whereas the parents' influence has diminished. In the past, physicians and parents alike were often spared from having to decide whether or not to sustain a severely impaired child's life. But with improved technology, physicians tend to do everything possible to prolong life regardless of its quality or its effect on the parents. In part, this is an adaptation to living in a litigious society, in which not using all available technology may be held to constitute a torte, yet this adaption is also another step away from a humane consideration of life.

Social inequities are also relevant to discussions of commercial surrogacy. In commercial surrogacy arrangements, a man with money pays a woman with less money for a service his wife cannot provide. Gender and money are not the only determinants of the distribution of power. Public opinion and the courts each have been used effectively to change the distribution of power in decision making.

Each of the contributors to this book closely examines decision making and reproductive choices and their consequences. Together they make a powerful statement about the existence of serious ethical dilemmas, and they begin the process of describing major problems in the making of moral choices in research and treatment involving reproductive technologies.

PART ONE

Ethics of Quality, Access, and Care During Pregnancy

1

The Ultimate Failure: Court-Ordered Cesarean Section

Brigitte Jordan and Susan L. Irwin

Throughout history and across cultures, birth has almost always been the province of women; yet Western biomedicine has increasingly colonized and technologized the birth event. The province of birth, formerly occupied by women and run with women's expertise and knowledge, has been taken over by biomedicine. This is evident in the transitions from home birth to hospital birth; from the involvement of community-based midwives to that of hospital-based obstetricians; from the use of everyday tools (such as scissors for cutting the cord and the marital bed for delivery) to high-tech equipment and procedures (such as ultrasound, fetal monitoring, cesarean section, and pharmaecological management). In the United States, several disturbing trends have come together that further medical and legal control of birth. The rate of cesarean sections has been steadily increasing so that more than 20 percent of all U.S. deliveries are performed by section (Gleicher 1984). At the same time, decisions about the management of pregnancy and labor are increasingly made on the basis of machine output (particularly electronic fetal monitors and ultrasound), in spite of considerable evidence that such data are open to varied interpretation. These trends, combined with the recent push for fetal rights, have set the stage for a relatively new phenomenon—the court-ordered cesarean section.

Rumors of court-ordered sections abound, but documentation is difficult to locate. Most of these cases are decided in local courts; as a consequence, the rulings are only available through local court records. Only when they are regarded as precedent-setting are local cases published, and only cases that come before state appellate courts are usually indexed and readily retrievable. Occasionally, cases have been described in journals and local newspapers. In addition, we obtained limited interviews with some of the people involved.[1] Altogether, we have information on six cases.

In a typical case, a woman, either close to term or in labor, is advised by a physician that a section is necessary. If the woman refuses, the physician, usually together with hospital administration and hospital lawyers, asks a local judge for a ruling. The unborn fetus is generally declared a ward of the court or some other agency and the mother is ordered to submit to surgery to save the fetus.[2] The legal and medical reasoning appears consistently to support the protection of a viable fetus over the rights and welfare of the woman. Among the cases we describe, two women gave birth vaginally; in one case, the manner of birth cannot be determined from the records; and in three cases sections were actually carried out. Legal judgments were obtained because it was believed that the life of the fetus was in jeopardy, but, in retrospect, no clearly negative outcomes justify these sections.[3]

Cases

Colorado, 1979

In the first case, a section was advised three and a half hours after the woman's membranes had ruptured, because of meconium-stained amniotic fluid, electronic fetal monitor data suggestive of fetal distress, high station (the baby has not properly descended in the birth canal), and failure to progress. The woman was described as angry, uncooperative, and obese. She refused to consent to a section, indicating fear of surgery (not an unreasonable fear given that she weighed over 300 pounds). Her family and a hospital lawyer could not persuade her to change her mind. After a psychiatrist judged her neither delusional nor mentally incompetent, the hospital sought a juvenile court order finding the fetus dependent and neglected and ordering a section. At the judge's request, a hearing was held in the woman's hospital room with court-appointed attorneys representing both the mother and the fetus. The court ruled in favor of the hospital, and surgery was performed eleven hours after admission. The baby was reported to be healthy; the initial Apgar score was 2 but the five-minute Apgar score was 8. (Named after pediatrician Virginia Apgar, this score reflects an assessment of the physical state of the newborn. A perfect Apgar score is 10; anything below 7 is considered indicative of fetal distress.) Although the initial low Apgar score reflected some fetal distress, the second score was in the range of normal for a newborn. The woman, on the other hand, suffered from delayed healing of the incision wound.

Georgia, 1981

One week before the expected birth, this woman was diagnosed as having a complete placenta previa (placenta covering the cervix). She refused a section on religious grounds. The hospital requested a court order authorizing the section as well as blood transfusions if necessary. The country Superior Court conducted an emergency hearing at the hospital at which a physician testified that there was a 99 percent probability of fetal death and a 50 percent probability of maternal death unless the section was performed. At the Superior Court's request, Family and Children's Services sought temporary custody of the fetus through the juvenile court. At a joint hearing, the courts ordered treatment and assigned temporary custody to the county Family and Children's Services. The parents petitioned the Georgia Supreme Court to stay the order, but their motion was denied on the same day. A third sonogram showed that the placenta had "migrated," and a few days later the woman gave birth vaginally to a healthy baby.

Illinois, 1982

In this case a section was recommended because of three prior sections, maternal anemia, and cephalo-pelvic disproportion (in which the baby's head is too large to fit through the maternal pelvis). The woman refused a cesarean section for religious reasons, and her husband supported her decision. A juvenile court judge ruled that the fetus was suffering medical neglect and awarded temporary protective custody to a hospital lawyer along with the power to consent to a section and to other medical or surgical procedures. We do not know if a section was actually performed. After the birth of a six-pound baby, custody reverted to the parents.

Michigan, 1982

This woman also had a diagnosis of placenta previa some weeks before the expected date of birth and, like the woman in Georgia, refused surgery on religious grounds. The hospital petitioned the county court, which, acting on the information that there was a 90 percent risk of fetal death, made the fetus a temporary ward of the court and ordered the woman to enter the hospital for necessary treatment. The woman went into hiding with her family and the police were unable to deliver the court order in spite of repeated attempts to locate her. She gave birth to a healthy baby vaginally three weeks later at another hospital.

Michigan, 1983

This case involved a West African woman whose first child was born vaginally after a section had been recommended and refused. In this particular labor, section was advised four hours after admission because of "secondary arrest [labor has stopped after it was well established] with failure to progress," based on a cervical dilation of 5 centimeters two and four hours after admission. Fetal heart tones were normal at the time, though earlier, late decelerations had been noted. The woman and her husband refused to consent to surgery. It appears that an administrator contacted a local circuit judge who expressed his willingness to order a section. During the legal process, the woman gave birth vaginally to a healthy child with Apgars of 8 and 9. The couple were unaware of the legal maneuvers at the time and, as far as we know, were never informed.

Michigan, 1986

A twenty-four-year-old, single, black woman, expecting her first baby, had had little prenatal care because she felt she was not getting proper attention at the prenatal clinic. She was admitted in early labor, but her contractions stopped and an induction was performed. After several hours, the woman was in great pain. Demerol was given, followed by an epidural (spinal anesthesia) several hours later, which apparently did not take. The patient was described as uncooperative and noncompliant, screaming with pain and thrashing about. A second epidural brought some relief, but "severe decelerations" in the fetal heart rate were noted in the chart. The woman was moved to the delivery room for a section, which she refused. At that time she was described as acting crazy, flailing her arms to keep the staff away. She was told that her baby could die or would have cerebral palsy, if it were born alive. The staff yelled at her and were ready to put her under when the hospital lawyers obtained a verbal okay from a local judge for the action. At that point, she was told that cerebral palsy meant mental retardation, and she is reported to have consented to the section with a whispered yes.[4] Surgery resulted in the birth of a 7-pound, 15-ounce baby with Apgar scores of 8 and 9.

Limitations of the Data

These cases raise a number of questions, some of which are unanswerable because the data are sketchy. Have any courts refused to order a section and on what grounds?[5] How widespread is this practice? Although lay persons think that court-ordered sections are rare, some professionals believe that there have been thousands of such cases across the country. Furthermore, it is unusually difficult to get detailed, accurate information even when we know that a court order was sought. In none of the cases we describe has it been possible to interview all participants, and even when we have had limited access to medical charts, it was difficult to accurately reconstruct events. But despite limited information, some general considerations and issues emerge.

Use of Force

The most striking feature of these cases is the use of legally sanctioned force. The use of force in obstetrics is not unknown, but the shifting of responsibility to the courts is new. A court order is a powerful device; judging from the cases described, once a court order is imminent, most hospitalized women in labor give up. The situation is different for a pregnant woman who is not yet in the hospital. She is able to generate support from family and community and implement strategies, the most effective of which is to remove herself from the reach of the medico-legal system. Once on medical territory, the potential for force, legally sanctioned or not, is pervasive. In at least one case, the staff were considering proceeding with the section even without a court order. Yet, the idea of physically forcing a struggling woman onto a section table is repulsive to most people, including medical personnel. On the legal side, it is worth noting that no court has ever forced a woman to submit to surgery to save the life of a child—as, for example, when the mother is the best match for a bone-marrow or organ transplant (Annas 1982).

Medical Justification

Most of the evidence on the basis of which court orders were sought for this group of women came from two sources: ultrasound for diagnosis of placenta previa; and fetal monitoring to assess the status of the fetus. In the two cases involving placenta previa, there was no placenta previa at the time of birth, and unless one believes in miracles ("God moved it," one woman said), one has to assume that the interpretations of the ultrasounds were in error.

In the cases in which a section was ordered because the fetus was in distress and threatened to die, the babies were born in good condition several hours later. Either babies can tolerate a lot more stress than is commonly assumed, or else the fetal monitoring data were unreliable, invalid, or misinterpreted. In the few randomized clinical trials that have been conducted, fetal monitoring has not been shown to improve outcome statistics (Banta and Thacker 1979). However, decisions to order a cesarean section involve more than overzealous interpretation of ambiguous machine recordings. In fact, the evidence for fetal distress or arrest of labor was in no case apparent from the record. The decision to do a section appears to be made independently of—and sometimes well in advance of—signs of serious medical problems. For example, six and a half hours after admission, one patient signed a consent to section (later revoked) because of fetal distress, though fetal heart tones had been entirely normal throughout the period. In another, "arrest of labor" was cited, though the woman had been making, in the words of another physician on the scene, "slow but steady progress." In every case in which we have interview data, there were medical opinions that conflicted with the assessment that a section was necessary. What, then, happened in these cases? Why were these women threatened with court orders when the medical justification was not at all clear cut?

The Portrayal of Women

The portrayal of the women who were the targets of court-ordered sections is strikingly negative. There is not a positive word said about any one of them in any of the court or medical records. As a group, they appear to be poor and ethnically diverse, sometimes single, uneducated, and without fluent command of English; some belong to marginal religious groups. They are not part of mainstream U.S. society.

Why these women refused a section is not easily determined. The officially cited reasons are those determined and noted by medical personnel and tend to be formulated within medical discourse. For example, in the case of the West African woman, it was noted that she refused because once she returned to her native country there would be no facilities for repeat sections; she was said to be willing to "sacrifice" this child in order to maintain her future fertility. It is much more likely that this woman relied on prior experience, having given birth vaginally with her previous child after a section had been recommended. The cited reason (which appeared for the first time in a case summary written some months after the birth) constitutes a medicalization of her thinking, for which there is no evidence in the information available to us.

Religion as a reason for refusal, which on first blush appears to fall completely outside the domain of medical discourse, is actually something with which the medical system has had a great deal of experience. Most hospitals have routine methods for getting court orders in the cases of parents who refuse blood transfusions for their children for religious reasons.[6] Thus, there is an established medico-legal procedure that can be used to deal with cases of this sort.

"Reasons" tend to be culturally valid justifications for a course of action, and so we are not surprised that the reasons heard and noted by the medical staff are medico-legal reasons. These may or may not reflect what motivated the women. They do, however, provide insights into what counts as justification for the events that transpired. In cases in which the woman's rationale is discussed, the assumption seems to be that if she were in her right mind, she would accept the medical recommendation. If she doesn't, it must be the result of ignorance, prejudice, unjustified fear of surgery, ominous "ulterior motives,"[7] or just plain craziness. What we do not find, by and large, is a willingness to consider the woman's position as reasonable or legitimate.

The Medical Staff's Predicament

To disallow the woman's perspective not only devalues the woman, it also confronts the medical staff with a predicament. Medical personnel see a "selfish and unreasonable" mother and are hard put to care for her in a nonjudgmental way. This difficulty clearly constitutes an ethical dilemma for those involved. The staff are obliged to care about the woman and for her, yet they find her position untenable; she appears irrational. From the information we have, it is clear that these have been very upsetting cases for the staffs. Particularly in the first and last cases, staff members had no sympathy for the women, yet they report that they should. There is evidence in at least some cases (such as chart entries in many different handwritings) that continuity of care and caring was lacking. Nobody wanted to take care of these women, who were perceived as troublemakers or worse.

At the same time, staff members sometimes disagreed with the diagnosis, with the decision to seek a court order, and with the potential necessity to apply force. In particular, nurses found themselves in a precarious position if they sympathized with the woman. They found it difficult to witness treatment they considered violative but were obliged to carry out doctors' and administrators' orders. In one of the cases, in which the woman gave birth vaginally before the order for a section was actually obtained, the staff thought that, even if there were a court order, no one would collaborate in doing the section.

Authoritative Knowledge and the Medical Construction of Reality

We know that in all obstetric systems some kinds of knowledge are considered important and consequential for the management of birth whereas other kinds are seen as irrelevant. Because childbirth in the United States falls into the medical domain, it is generally assumed that decisions about its management are to be made by physicians who, drawing on medical technology for information, are seen as uniquely and exclusively qualified to judge and predict the course of labor. The medical model of birth is thus a technology-based model, in which those who control the technology hold "authoritative knowledge," whereas whatever information, experience, or conviction others may bring to the birth process is specifically disallowed.

Most members of society subscribe to this technomedical view of birth, including childbearing women, medical professionals, lay persons, lawyers, and judges. In cases of the sort we have been discussing, there comes a point when the judge asks the physician in charge whether, to the best of his knowledge, the fetus will die without the section but can be saved if surgery is performed. Judges ask this question of physicians because they, like everybody else, see the physician as holding "authoritative knowledge"—that is, the crucial knowledge that will and should determine the management of the birth. The power of this model is such that no other views need be considered, as only the physician has access to the technology that generates legitimate information.

The birth process itself constitutes a rite of passage in which society's basic values, centering on science, technology, and the power of institutions, are reaffirmed and, if need be, enforced. It conveys to birthing women both the basic values of society and the technological belief system that underlies these values. As society's representative, the American obstetrical system is instrumental in instilling these values into birthing women during their individual passages into motherhood (Davis-Floyd 1986). Most women willingly submit themselves to the authority of the medical view without questions. They manage to experience the technologies and procedures as reassuring and the delegation of authority to physician as functioning in their own (the women's) best interests.

For most women, the technomedical approach to birth is of a piece with their prenatal care, in which monitoring of the progress of their pregnancy also relied on sophisticated obstetric technology.[8] They are thus well prepared for the implementation of such technology in the hospital, expect it, and often even prefer it (Jordan 1983). The women who consciously reject the technomedical model of birth never appear in the hospital at all, preferring to have their babies at home. These women tend to be educated and tend to have an extensive support system for their choice. On occasion, however, a woman comes into the system who has not internalized the technomedical model of birth, be it for cultural, religious, or other reasons. It is these women—entering the system without information, financial resources, or support—who become the victims of court-ordered sections. In the cases we have described, women have refused to go along with the dominant cultural construction. And, as is to be expected, those in control take steps to bring them into line. The challenge that they present to the dominant construction of reality is overridden by recourse to another powerful institution in this society—the court system.

Informed Consent

The legal notion of informed consent requires that a patient be given information regarding his or her condition, including treatment options; that an effort be made to ascertain that they understand what they have been told; and that patients make their decisions without coercion or duress.

As Lidz et al. (1983) have pointed out, this doctrine fails to deal with the realities of medical work and interaction in medical settings. Informed consent sees the patient as the decision-maker, whereas in practice it is the physician who has already made the decision when informed consent is sought. This was certainly what happened in all of the cases we describe. Furthermore, there seems to have occurred a fundamental failure in communication, a failure resulting from the fact that the women's views were never considered legitimate. Thus, there was no negotiation process that could have brought the differing versions of reality into accommodation (Kleinman, Eisenberg, and Good 1979). Nevertheless, it would be a mistake to see the court orders simply as the result of miscommunication. Rather, they have to do with an unresolved conflict between the patient-centered meaning of informed consent and the nature of medical practice, in which decision making is hierarchicized and the patient is the object of decisions, not an active participant.

We take it to be self-evident that informed consent must include the possibility of informed refusal (Hahn 1982). Some decision-making power must reside in the patient in order to speak meaningfully of informed consent. Otherwise, patients are simply certifying that they have been told what is going to be done to them. In at least some of the cases, it appears that providers made some effort to inform the people concerned, who seemed to understand what they were told, and, after weighing this information, decided that they did not want the section. It is true that the outcome of these exchanges was not to the liking of or in accordance with the professional judgment of the providers. Nevertheless, the process of requesting consent implies that a refusal will be respected. Taking steps to obtain a court order for a section negates the negotiation process between patient and provider by attempting to make sure that only one kind of outcome will obtain—namely, the section.[9]

Conclusion

What we see here is an official, socially validated construction that opposes the interests of the mother to those of the fetus. This construction is natural within a "rights" approach, whereby the rights of the fetus are pitted against the rights of the mother. Overwhelmingly, this has been the language in which this debate has been conducted. We would suggest, however, that this is not the way in which women are likely to think about these issues. Feminist scholars exploring women's "ways of knowing" (Belenky et al. 1986) have found that women, much more than men, find the foundations for their thinking and the metaphors for their being in the world through intimate relationships rather than hierarchical ordering; in close connection rather than objective distance; in the particular rather than the abstract; in understanding rather than control. Yet this voice is nowhere audible—there is no place for it in the din generated by the authoritative clamor of medico-legal argumentation.

At least some of these women made moral judgments and decisions not by thinking in adversarial ("rights") terms but by considering relationships. For the women whose decisions and fates we have discussed, the relevant relationships include not only the one between this woman and this fetus but also the woman's responsibilities to herself, to existing children, her partner, her social network, potential future children, and more. To argue fetal versus maternal rights may be to speak in a voice that is not the voice of women.

Acknowledgments

We thank Mary Alfano, Robbie Davis-Floyd, Tracy Dobson, Robert Hahn, Bruce Miller, Charles Senger, Ronald C. Simons, Barrie Thorne, and Mary Ann Zettelmaier for discussion and constructive comments on earlier drafts. We also want to express our deep gratitude to the individuals who were willing to talk to us about their involvement in court-ordered cesarean sections.

Notes

1. Such cases will not be identified in order to protect the anonymity of informants.

2. A seventh case differs dramatically from these six. A sixteen-year-old Southeast Asian woman who spoke almost no English was admitted in labor with a breech fetus. She was accompanied by a group of family members who, led by her father, refused to permit the section that was advised when she began showing signs of preeclampsia (a condition that can lead to potentially fatal convulsions, promptly relieved by delivery of the fetus). In this instance, the court order was sought for the benefit of the mother, not the fetus. The physician reported that he would not seek a court order for purely fetal indications. Since writing this chapter, we have collected half a dozen more cases, some of which are discussed in Irwin and Jordan 1987:323–325.

3. There were no medically negative outcomes for the fetuses involved; in other words, the status of the babies after birth, as determined by Apgar scores and medical reports, was good. There is little information on the medical or psychological effects on the women, though in one case it was noted that she suffered delayed healing of the incision.

4. It has been suggested that the woman, in such cases, may hope to be freed of an unwanted pregnancy (Cf. Leiberman et al. 1979).

5. Late in 1987, St. Vincent's Hospital asked New York City Civil Court Judge Margaret Taylor to order a section for a thirty-five-year-old indigent woman who had borne ten children and had seen many women in her neighborhood suffer

complications from obstetric and gynecological surgeries. The hospital said that a section was necessary because the umbilical cord was wrapped around the neck of the fetus. Judge Taylor ruled that there was no legal basis for forcing a pregnant woman to undergo treatment to benefit a third party that any other legally competent man or woman would be allowed to refuse. Several hours later, the woman gave birth to a healthy child (Lewin 1987).

6. We have some information from interview data that suggests that the working relationships and mutual understandings produced through cases involving Jehovah's Witnesses provide the channels for obtaining court orders for cesarean sections.

7. See note 4.

8. The powerful socializing effect of prenatal care may be one of the reasons why women who have had none are so despised by medical staff. The label "a walk-in" carries with it a moral condemnation that goes beyond the recognition of potential medical problems that might have been detected by prenatal care. What it signifies is that such a woman has not been properly socialized and may be a troublemaker.

9. This raises the question of who the appropriate parties to the consent/refusal negotiation are. Is it physicians and patients, or are there others who have an overriding interest? It appears that in some, if not all, of these cases, hospital administration was substantially involved.

References

Annas, George J. 1982. Forced Cesareans: The Most Unkindest Cut of All. Hastings Center Report 12:16–17.

Banta, David, and Stephen Thacker. 1979. The Premature Delivery of Medical Technology: A Case Report. Washington, D.C.: National Center for Health Services Research, Department of Health, Education, and Welfare.

Belenky, Mary Field; Blythe McVicker Clinchy; Nancy Rule Goldberger; and Jill Mattuck Tarule. 1986. Women's Ways of Knowing: The Development of Self, Voice, and Mind. New York: Basic Books.

Davis-Floyd, Robbie. 1986. Birth as an American Rite of Passage. Ph.D. dissertation, University of Texas at Austin.

Gleicher, Norman. 1984. Cesarean Section Rates in the United States: The Short-term Failure of the National Consensus Development Conference in 1980. Journal of the American Medical Association 252:3273–3278.

Hahn, Robert A. 1982. Culture and Informed Consent: An Anthropological Perspective. Making Health Care Decisions, Vol. 3. Washington, D.C.: President's Commission for the Study of Ethical Problems in Medicine and Biomedical and Behavioral Research.

Irwin, Susan, and Brigitte Jordan. 1987. Knowledge, Practice and Power: Court-Ordered Cesarean Sections. Medical Anthropology Quarterly 1, no. 3:319–334.

Jordan, Brigitte. 1983. Birth in Four Cultures: A Crosscultural Investigation of Childbirth in Yucatan, Holland, Sweden and the United States. 3d edition. Montreal: Eden Press.

Kleinman, Arthur, Leon Eisenberg, and Byron Good. 1979. Culture, Illness, and Care: Clinical Lessons from Anthropologic and Cross-Cultural Research. Annals of Internal Medicine 88, no. 2:252–258.

Leiberman, J. R., et al. 1979. The Fetal Right to Live. Obstetrics and Gynecology 53:515–517.

Lewin, Tamar. 1987. Courts Acting to Force Care of the Unborn. New York Times, November 23, pp. 1, 12.

Lidz, Charles, et al. 1983. Barriers to Informed Consent. Annals of Internal Medicine 99:539–543.

2

Chromosomes and Communication: The Discourse of Genetic Counseling

Rayna Rapp

In 1958, Jerome LeJeune peered through a microscope and identified an extra chromosome (the forty-seventh) in samples of smooth connective tissue taken from three patients with Down's syndrome. In the late 1960s, several teams of doctors reported the presence of the same extra chromosome in amniotic fluid extracted from pregnant women's uteruses. These discoveries suggested possibilities for the prenatal diagnosis of Down's syndrome, as well as for other inherited disabilities.

The practice of amniocentesis for prenatal diagnosis of inherited disabilities developed as the ability to identify, stain, and study chromosomes increased. At the same time, the DNA (deoxyribonucleic acid) revolution transformed the fields of cytogenetics and cell biology, and by the late 1980s more than two hundred inherited conditions, most of them extremely rare autosomal recessive diseases, could be diagnosed prenatally. Research frontiers in human genetics and molecular biology became inextricably tied to the routinization of amniocentesis.

The rapid proliferation of information about human genetics had implications not only for health workers in the field of reproductive medicine but for pregnant women and their families as well. The profession of genetic counseling was developed to aid the public in translating the discourse of human genetics, with its technical and ever-shifting implications, into usable and more popular language. In this chapter I will explore the social impact and cultural interpretations of prenatal diagnosis in the context of genetic counseling. During two years of fieldwork in New York City I accompanied a team of genetic counselors on hospital rounds, observed a cytogenetic laboratory, and interviewed thirty-five genetic counselors, numerous pregnant women, and their families. Throughout my analysis, I implicitly describe the paradox of genetic counseling: Developed to provide pregnant women and their families with more choices, it also inadvertently replicates and extends the social hierarchies that limit reproductive choices.

The Genetic Counselor

In 1969, program developers at Sarah Lawrence College in Bronxville, New York, planned a new science masters degree program in the School of Continuing Edication. In 1971, the first formally trained genetic counselors graduated. The term *genetic counseling*, coined in 1947, initially implied an ethical-neutral process that favored personal choice in the century-old eugenics debate about society's responsibility and right to encourage or discourage reproduction in certain individuals and families (Kevles 1985; Reed 1974). Before modern genetic counselors appeared on the scene, research pediatricians, geneticists, and immune biologists—the majority of whom were men—would counsel families with genetically disabled members about recurrence risks and disease management. Once amniocentesis became commonly available, there was a need for someone to explain the risks and benefits of the test and to translate scientific possibilities into personal terms. As it turned out, women seemed drawn to the role of "gatekeeper" between science and social work, between epidemiology and empathy. The attraction may have been based on common experiences of pregnancy, childbirth, and nurturing and on an interest in counseling. Thus, the first wave of genetic counselors tended to be well-educated women who lived near Sarah Lawrence; many were married to professionals. Many of these women had raised their children and were interested in returning to school. They brought with them a specific set of upper-middle-class family values. The program promised the students an amalgam of hard science and counseling skills and held out the possibility of part-time work. The study of advances in genetic science provided an exciting avenue back into the workforce.

Genetic counseling quickly became a profession-in-formation (Rollnick 1984). Sarah Lawrence's program is still the largest and perhaps the most respected one in the United States, but there are now approximately nine others (depending on how one defines "program") located in seven states, the District of Columbia, and Canada. Although the curricula vary somewhat, all the programs provide training in human genetics, medical genetics, and counseling and include supervised clinical internships. Some programs include a seminar on bioethics, or ethical issues in genetic counseling, which has a strong orientation toward individual choice. But according to the counselors I interviewed, there is rarely much discussion of the cultural constraints and resources within which pregnant women and their families may be operating. Most programs require two years of study. Board certification for practitioners is under way, and genetic counselors are insured by the medical centers for which they work. The National Society of Genetic Counselors (NSGC) estimates that there are currently six to seven hundred practicing genetic counselors, including not only those who have completed a two-year master of science program and served supervised internships but those who moved into the field from nursing, social work, and related disciplines.

Genetic counseling remains a "women's field." Less than 1 percent of the graduates of genetic counseling programs are men, and many of these are employed in administration. Current students and recent graduates of genetic counseling programs are less likely to have attended elite women's colleges, are more likely to have considered and rejected a premed track, and are somewhat more diverse in class, ethnic, and racial background than the first wave of Sarah Lawrence graduates. Although consciousness of minority issues in counseling is growing, the numbers of NSGC counselors from Black-American, Hispanic, or Asian-American backgrounds remains small—not more than 5 percent.[1]

As part of an ongoing study of the social impact and cultural meaning of prenatal diagnosis, for over two years, one to two days a week, I observed the counseling sessions of five genetic counselors working for New York City's Department of Health. As an observer and, sometimes, as participant-translator, I sat in on more than two hundred intake interviews and had extensive informal conversations with the five counselors. I also interviewed thirty other genetic counselors who work in New York City, at least one from each medical center that offers amniocentesis. Although most counselors in my sample were white and middle class, at least five spoke Spanish as their first language, and one was Black-American.

In New York City, unlike many other parts of the country, a combination of state and city policies both fund and offer prenatal diagnosis to women regardless of their ability to pay for the test themselves. The City Laboratory (through which my fieldwork was conducted) collects amniotic fluid samples drawn from a population of pregnant women approximately one-third Hispanic, one-third Black-American, and one-third white. The city lab has contracts to perform cytogenetic analyses for twenty-four urban hospitals in all boroughs and neighborhoods, encompassing both municipal and voluntary facilities. Women seen by the city's genetic counselors are roughly 50 percent private patients and 50 percent clinic patients (that is, poor and working class). These genetic counselors are "circuit riders": In the period of my observations, they rotated among six to eight hospitals—one serving only private patients, four serving an exclusively clinic population, and several serving both. Genetic counselors must thus explain the test to a diversity of patients and may help to shape experiences of knowledge and power in the reproductive lives of many women whose backgrounds are significantly different than their own.

The genetic counselor tends to be highly self-conscious of the ethical and scientific roots of her new profession. The counselors I interviewed were very much aware of the anxiety, as well as the relief, that their services invoke. Most had thought deeply about why someone might reject, as well as accept, amniocentesis and the subsequent possibility of abortion. And many were curious and insightful about the problems of cross-cultural communication.

The Discourse of Genetic Counseling

A genetic counseling session almost always precedes the use (or nonuse) of amniocentesis. Counselors meet with their pregnant patients (and any supporters the woman brings with her) in the hospital where the test is offered. In course of an hour's visit, counselors convey a great deal of medico-scientific information, ask and answer questions, and prepare women to take the test. The interaction is conversational, but the discourse of genetic counseling is resolutely medico-scientific, revealing and creating some meanings, which mask or silence others. Medical language deploys a great authority and cannot always be used to respond to the resources and questions particular women bring to this encounter. Miscommunication as well as communication, silence as well as conversation, fill up a genetic counseling appointment.

Genetic counselors generally begin communication with pregnant patients with three goals in mind: to convey significant information about the risks of birth defects and the availability and nature of amniocentesis; to take a health and family history; to communicate with the patient well enough so that her questions and concerns can be addressed. In order to accomplish these tasks, most genetic counselors begin by "setting up a dialogue." Many begin intake sessions by posing some variant of the question, "Do you know why you are here talking to me?" Middle-class, scientifically educated pregnant patients may respond, "We're planning to have amniocentesis," thus pushing the counseling script into high gear. Implicit in that answer is not only a knowledge of appropriate medical indications for the test but the existence of "the couple" as a decision-making unit. But a Dominican mother of three may answer, "por culpa de mi edad" (literally, "for the fault of my age"), thus presenting the counselor with several options. She can assume that the woman *knows,* as *age* is the factor that sends her here. Or she can confront the *culpa* head-on, explaining that older women having babies present no shame, just medical risks. But wherever she begins her routine explanation, the counselor is likely to have to adjust her language to the language and assumptions of her pregnant patient (and sometimes the patient's mate).

In adjusting (or not adjusting) to the patient's language, genetic counselors are bound by the limits of their own communicative resources. Although many native languages are spoken by the pregnant women, most counseling sessions are conducted in English. But at least five counselors in New York hospitals are native Spanish speakers, and another five are comfortable counseling in Spanish. Many others have learned a bit of medical Spanish and work through translators—who range from trusted assistants who understand the counselors' agendas (a secretary in the office; a clinic nurse) to catch-as-catch-can interpreters (the ten-year-old child brought along by the patient; a husband, brother, or male neighbor, who may be embarrassed to find himself discussing prior miscarriages and abortions).

Availability of fluent translation is a significant problem: Depending on the hospital's catchment area, patients may be close to 100 percent Spanish-speaking, 50 percent French/Creole-speaking, or about 25 percent monolingual Spanish-speaking—situations I observed at various New York City hospitals. Native-language categories only approximate the variety of communicative differences that genetic counselors confront. "Hispanic" glosses a range of Spanish-speaking cultures, especially at the present time in New York City. Some genetic counselors distinguish Hispanics (often meaning Puerto Ricans and Dominicans—the "old" migrants) from new migrants, who might be "middle-class" Colombians and Ecuadorians or the "field mice" (the "poor, rural, and humble") of Central America. Although exact cultural differences among Spanish-speaking groups may be unknown, most counselors recognize something of the diversity in educational levels, familiarity with medical terminology, and religious observation that different nationalities may represent.

Nominal or deep fluency in another tongue may aid, but does not insure, direct communication for science-speakers. Language differences may signal communicative ambiguities far beyond the question of literal translation. Local metaphors of pregnancy, birth, and parenthood do not necessarily translate easily into the realm of medical discourse. Two native Spanish-speaking counselors pointed out the far-reaching impact of their conversations with pregnant patients:

> This knowledge is more than genetic. They learn about things that were completely hidden, where the eggs are, what sperm does, how children get to look like their parents. They have ideas, but this is female physiology, it is knowledge, not just information. For this, they come back.

> When I see confusion, I go to work, I tell them in language they will understand, language of the streets. They are comfortable here, it is a good place to visit, they come back to see me whenever they come to the hospital.

Of course, not everyone is equally open to the complex relation between native tongue, knowledge, and communicative power. Two counselors, one of whom conducts group sessions for patients in Spanish, expressed irritation that so few of their clients "bothered to learn English": "They're here ten, maybe fifteen years. They learn enough English to pick up their welfare checks. Why don't they just learn the language? My grandparents did."

And if Spanish presents difficulties, French and Haitian Creole are virtual *terrae incognitae*. Only one genetic counselor feels comfortable counseling in French; none knows any Creole. The lack is significant: In at least one city hospital, Haitians make up about 50 percent of the current patients referred for counseling. In translating for counseling sessions, I discovered that recent immigrants coming from the Haitian countryside do not recognize Down's syndrome; no word exists in Creole for the condition. In principle, the incidence of Down's syndrome is invariant worldwide. But in a country with the worst infant mortality statistics in the Western hemisphere, babies may die from many causes, and this one may go unrecognized as a syndrome. Nonrecognition of the label may also reflect other cultural and political experiences. Haitians living in New York City already confront alternative definitions of their children's vulnerabilities. As one Haitian Evangelist father told me, while firmly rejecting amniocentesis on his wife's behalf, "What is this retarded? They always say that Haitian children are retarded in the public schools. But when we put them in the Haitian Academy (a community-based private school), they do just fine. I do not know what this retarded is." In his experience, "chromosomes" seem a weak and abstract explanation for the problems a Haitian child may face.

In addition, the language of genetic counseling is resolutely statistical; it is an axiom of good counseling that a patient must be told her risks before she can decide to take or refuse the test. Yet, statistics imply an abstract mathematical universe that may not be shared by patients who have little formal schooling. The majority of genetic counselors confront this problem by simplifying the numbers and adding information, if it is requested:

> [to someone perceived as unable to handle numbers]: At your age, the risk of having a baby with mongolism is about one in a hundred.
>
> [to someone perceived as uneducated but attentive]: Pregnant ladies your age have a 1 in 106 chance of having a baby with this condition. That means that of every 106 pregnant ladies your age, 105 will have no problems, and 1 will have a child with the problem.
>
> [to someone perceived as scientifically sophisticated]: At 35, a woman's risk of bearing a liveborn child with Down's syndrome is 1 in 385; at 40, it increases to 1 in 106; at 45, it is 1 in 30.

Likewise, miscarriage rates must be shared:

> This is a very safe test, but there's always *some* risk to any test in medicine. The risk of losing the baby after amnio is very small, but it isn't zero.

versus:

> Amnio adds three miscarriages per 1,000 to those having the test. Of 1,000 women your age sixteen weeks pregnant who don't have amniocentesis, 32 will not have a liveborn child at the end of the pregnancy, through miscarriage or stillbirth. Of 1,000 women who have the test, 35 will lose the pregnancy.

But code switching and simplification of numbers only mark the professional side of the interaction. Such strategies may sit comfortably with information-seeking, medically compliant patients, especially those with some advanced education (middle-class), but the numbers may still not have meaning for less-privileged women with different experiences. Low-income black women, for example, often had a radically different response to the statistics than did middle-class couples. When a woman has given birth to four other children, comes from a family of eight, and all her sisters and neighbors have had similar histories, she has seen scores of babies born without recognizable birth defects. It requires a leap of faith in abstract reasoning to accept that a number produced by a lady in a white coat means that the risk of having a baby with a birth defect is steadily rising with each pregnancy. Among middle-class professional families, in which childbearing is likely to be delayed, the pregnancy in question is probably a first or, at most, a second one, and children are likely to be scarce throughout the couple's network. To such a couple, 1 in 300 sounds like a large and present risk, whereas for the low-income mother of four, the same number may appear to signify a very distant and small risk.

From the vocabulary of biomedicine the counselor selects for concepts used to describe pregnancy and birth, abortion and disability. In a forty-five-minute intake interview, code switching occurs rapidly, as counselors feel out their clients: "babies" vie with "fetuses" for space in "wombs," "tummies," or "uteruses"; "waters" or "liquid" or "fluids" may be "taken out with a needle" or "withdrawn through an insertion"; the "test" or "procedure" may involve "looking at the inherited material" or "examining chromosomes." In the worst-case scenario, women must decide to "terminate" or "abort" an "affected" or "sick" pregnancy in which "Down's syndrome" or "mongolism" has been diagnosed. In the war between medical and popular language, the more distant idiom may provide reassurance by suggesting to some pregnant women that their experiences are part of medical routine (Brewster 1984). But for others, medical terminology may muffle anxiety-provoking choices until they are expressed through dramatic disruption: "So I was sittin' and listenin', listenin' and sittin' and all the time gettin' more and more preoccupied. The counselor kept on talkin' but she never did say it, so finally I had to just say it, right while she was still talkin': 'You can't take the baby out *then* [so late in pregnancy], can you now?' I finally asked" (Veronica, 36, Trinidadian-born factory worker).

Much of the scientific information that counselors want to convey exists in technical and invisible terrains. Most counselors work with visual aids, especially with clinic patients, attempting to map what patients cannot see for themselves. These may include charts, graphs, and karyotypes (pictures of chromosomes arranged in pairs, to illustrate females and males, normal and abnormal genotypes). Many counselors show pictures of children with Down's syndrome. And almost all discuss the sonogram accompanying amniocentesis, in which "you can see the baby moving around."

Developed to present the mysteries of the womb, the workings of heredity, and the universe of epidemiology graphically, such icons of professional knowledge are not, to be sure, self-evident. They require interpretation during which health professionals not only reveal some of their arcane wisdom but also shape the perceptions of the client:

> I saw the sonogram of the twins, and I was thrilled. But I really couldn't read it, I didn't know what it meant. They had to interpret it for you, to say, "here's a heart, these are arms." Afterwards, it made me queasy—they made the babies real for me by telling me what was there. If they hadn't interpreted, it would have just been grey blobs, and now, I'm more frightened to get the results of the amnio back. [Daphne, 41, U.S.-born professional]

> It was nothing, really, it looked like nothing. Then they showed it to me, and made it something. [Ileana, 37, Ecuadorian-born babysitter]

Virtually all counselors have a minimal "threshold" of information that they need to explain. For some, it includes the concept of chromosomes and genes, for others, it is the idea of heredity, especially hereditary health problems. For most, it is the 2–3 percent risk of a birth defect involved in any pregnancy and the increased risk of chromosomal problems associated with childbearing later in life. Whatever her personal goal, no counselor is satisfied if she feels that a patient has failed to grasp her minimal scientific scenario. But medically significant lessons may mask the social experiences and meanings that disabling conditions hold, a point to which I will return. Scientific discourse silences as well as articulates.

In their training, counselors are taught to illustrate chromosomal problems by reference to a forty-seventh chromosome, which clinically manifests as Down's syndrome and its accompanying mental retardation. Although this condition is almost universally recognized, the content of that recognition varies considerably. Many families share the counselor's concerns about the limitations on independence that mental retardation represents in our culture. But in families who have had direct experience with children with Down's, consciousness of these disabilities may be more finely honed: Down's children may be mildly, moderately, or profoundly retarded; may suffer from heart or esophagal problems, hearing loss, or increased risk of leukemia. "Mental retardation" provides an iconic description that blurs differences among Down's children, even as it categorizes them.

And counselors do not routinely offer information on the social realities, rather than the medical diagnoses, that accompany this (or any other) condition. Many counselors were open, in principle, to questions concerning support for families with genetically disabled members (especially if the counselors worked in pediatric as well as prenatal services, they may have often come into contact with such families). However, they had little or no experience in discussing either the social stigmas of having a Down's syndrome baby or the infant stimulation programs that would be available for such a newborn. Unless propelled by a patient's question, they did not transcend the medical framework of the discussion by offering a visit with a family whose child had the diagnosed condition about which the woman was concerned. Thus, the counseling session is likely to reinforce a medical, rather than a social, definition of the problems of childhood disability.

Indeed, the assumptions of medicine weigh heavily throughout an intake interview. Genetic counselors elicit health histories, using a standard questionnaire. From a counselor's point of view, recent immigrants, especially from very poor countries, are likely to exhibit shallow knowledge of their own heredity. The cause of a father's death, the name of an uncle's form of mental retardation may be known, especially if births and deaths occurred after the immigrant left home. And although some health experiences may be vividly recalled—high fevers, exposure to x-rays—others seem irrelevant, or badly named, from the migrant patients' points of view. Many Haitians, for example, routinely answer no to all questions concerning family histories of heart and kidney disease, diabetes, and venereal diseases. Their negative answers may well be ambiguous, as serious conditions may have gone unreported or unnamed because they were virtually untreatable for all but the most privileged elite. A community health outreach worker from Jamaica, now working in Brooklyn, told me this part of her life story: "Sickle cell, do I counsel sickle cell, sure I do now. But then, I didn't know what it was. My brother, he died of it back then, we didn't know, no one told us. What's the difference? No transfusions back then, anyhow." But to the counselor, the negative answer may result in one of a number of misinterpretations, such as that the family lacks health problems or the patient lacks interest or intelligence. Such interpretations are overly determined by the isolated nature of a medical health history, out of its epidemiological and cultural context.

Through claims of universality that silence other cultural resources and worldviews, the codes, genres, and assumptions of biomedicine limit the conversations genetic counselors may have with their patients. The language of biomedicine also limits communication by locking counselors into a discourse dominated by technical language, despite their sincere interests in reaching out to patients. Counselors are caught between the need to sound authoritative and the desire to "glide on the patient's wavelength," as one counselor described the situation. Genetic counselors, as new professionals, must lay claim to a monopoly on the information that they offer. Such a claim must impress the rest of the medical hierarchy, in whose ballgame they are new players.

At the same time, genetic counselors are extremely interested in the public image of their field, and many are eager to further public education in its issues. Individually, many enjoy speaking at popular health forums, being interviewed by journalists, and serving as consultants to media projects through which they hope scientific knowledge will be effectively disseminated. This tension between monopolizing an arcane body of information and popularizing it is inherent in a new profession. Members need to "sell their services" both to the medical establishment and to the individual clients, for whom they serve as translators (Brown 1986). But this tension cannot easily be resolved in favor of popular language: Medical discourse is authoritative, it pervades the hospital setting, and it claims universality. The languages of patients are polyglot and usually have no lexicon—or an unacceptable lexicon—for the expression of clinical and epidemiological facts.

This contradiction between speaking like doctors and talking with patients surfaces in nonverbal communication as well. Some counselors wear white coats "to appear like medical professionals," for example, whereas others forego that symbol of antiseptic separation, hoping to "make the patient feel right at home." One counselor described her dress code to me: "On clinic days, I like to dress comfortably, so they will feel at home, casual with me. I dress up like a professional when I see private patients; they expect more from their medical providers." The markers of professional status extend to briefcases, charts, and visual aids, isomorphic with those of both health-care professionals and their private clientele. But the poor carry markers with them, too: Non-English speakers often bring electricity bills, check stubs, or personal letters to communicate home addresses. And the ubiquitous Medicaid and clinic cards that define payment status and rights serve as passports into medical domains, marking client status.

Disability, Variability, and "Choice"

In a genetic counseling intake interview, rich, many-layered, and powerful messages are being communicated. Officially, information about a new reproductive technology—its risks, limitations, and possible benefits—is being conveyed. Unofficially, the power to define a pregnancy, fetus, disability, and maternal responsibility for fetal health are all under negotiation: basic knowledge about human heredity, reproduction, and its control simultaneously, as several of the Spanish-speaking counselors tried to indicate. Ex-officio, women (and their families) are being given a set of choices about the kinds of babies they might, or might not, accept to bear.

> I was hoping I'd never have to make the choice, to become responsible for choosing the kind of baby I'd get, the kind of baby we'd accept. But everyone—my doctor, my parents, my friends—everyone urged me to come for genetic counseling and have amniocentesis. Now, I guess I'm having a modern baby. And they all told me I'd feel more in control. But in some ways, I feel less in control. Oh, it's still my baby, but only if it's good enough to be our baby, if you see what I mean. [Nancy, 36, U.S.-born lawyer]

Such choices are, of course, far more than individual, for they emerge from the embeddedness of specific pregnancies, contextualized by ethnic, class, racial, religious, and familial experiences. Thus, for Creole-speaking Haitian immigrants from the countryside who may not recognize Down's syndrome, being "offered the test" for a condition that holds no cultural meaning may make no sense. Recognition of a medical condition may be contextualized within other powerful discourses. For example, an Ecuadorian evangelical and a Colombian Catholic, both opposed to abortion but still awaiting results of the test, made similar points to me: "Science can reveal God's miracles, let you know what He has in store." Their desire for the test might be interpreted as a bridging of religious and secular-scientific cultures, an expression of the necessities and possibilities of living in a multicultural world.

Even for Black-Americans raised in the shadow of medical discourse, the choices may seem inappropriate: One pregnant black woman who was adamant about her antiabortion stance wanted the test in order to know whether she ought to move back to Georgia, where her mother would help her to raise a disabled child. Another, who had recently given birth to a baby with Down's syndrome, said to the genetic counselor, "My kid's got a heart problem. Let me deal with that first, then I'll figure out what this Down's business means." And low-income Puerto Rican parents I interviewed at an infant stimulation program said of their daughter with

Down's syndrome, "She's growing really well, we were only concerned that she wouldn't grow, that she'd be really small. But now that she can walk, and she's growing, she seems like a normal child to us." Mental retardation is the key focus when genetic counselors speak about Down's syndrome and offer amniocentesis, but it may not be the most significant factor in the consciousness and decisions of many of their clients. Among the Spanish speakers I interviewed, physical vulnerability, especially if it was highly visible, seemed a much more urgent problem for family life. The choice any pregnant woman makes to take or reject the test and to keep or end any specific pregnancy flows from the way the pregnancy is embedded in the totality of her life.

If Down's syndrome represents the iconic case for genetic counseling, other potential diagnoses are more ambiguous and present complex problems for communication. The sex chromosome anomalies (XXY, or Klinefelter's syndrome; XXY, XXX, and Turner's syndrome, or XO) all spell problems with growth and fertility, but none is "incompatible with life," as medicine would express it. Controversial claims concerning mental retardation, learning disabilities, and, in the worse case, antisocial behavior, are associated with these conditions, but all are contested because there is no baseline population from which to make scientific comparisons. Only people who are diagnosed as having a clinical problem will ever have their chromosomes "read." And even when the ethical complexities of collecting baseline data on anomalous sex chromosomes are sorted out, epidemiological patterns cannot predict whether affected individuals will express many symptoms of the condition and whether clinical expression will be severe or mild.

So, when one of these diagnoses is made based on a test tube of fetal cells, its meaning is open to interpretation. One genetic counselor encountered two patients, each of whom chose to abort a fetus after learning that it had XXY sex chromosomes (Klinefelter's syndrome). One professional couple told her, "If he can't grow up to have a shot at becoming the president, we don't want him." A low-income family said of the same condition, "A baby will have to face so many problems in this world, it isn't fair to add this one to the burdens he'll have." And a Puerto Rican single mother who chose to continue a pregnancy after getting a prenatal diagnosis of Klinefelter's said of her now four-year-old son, "He's normal, he's growing up normal. As long as there's nothing wrong that shows, he isn't blind or deaf or crippled, he's normal as far as I'm concerned. And if anything happens later, I'll be there for him, as long as he's normal-looking."

From a patient's point of view, most diagnoses are inherently ambiguous.[2] An extra chromosome spells out the diagnosis of Down's syndrome, but it does not distinguish mildly from severely retarded children or indicate whether this particular fetus will need open-heart surgery. A missing X-chromosome indicates a Turner's syndrome female but cannot speak to the meaning of fertility in the particular family into which she may be born. Homozygous status for the sickle-cell gene cannot predict the severity of anemia a particular child will develop. All such diagnoses are interpreted in light of prior reproductive histories, community values, and aspirations that particular women and their families hold for the pregnancy being examined.

Values, Decisions, and Power in Genetic Counseling

The ethical complexity of diagnoses is something all genetic counselors confront. Counselors are trained to be empathic as they convey statistics, and to practice Rogerian therapy—that is, a therapeutic style that is noninterventionist, aimed at helping the patient to make up her own mind. This counseling model assumes that the professional and the patient mutually participate in a decision-making process. Their task is short term and well defined: Should the woman accept amniocentesis? If a positive diagnosis is made, should she end or continue the pregnancy? Unlike a more directive model (which counselors associate with physicians), this protocol assumes that the patient can and must decide for herself. Yet, counselors all know how hard it is to keep their own feelings out of a given situation.

> We're suppose to ooze empathy, but stay aloof from decisions.
>
> Oh, I know I'm supposed to be value-free. But when you see a woman on welfare having a third baby with one more man who's not gonna support her, and the fetus has sickle-cell anemia, it's hard not to steer her toward an abortion. What does she need this added problem for, I'm thinking?
>
> So I try to put it in neutral, to go where she goes, to support her whatever her decision. But I know she knows I've got an opinion, and it's hard not to answer when she asks me what I'd do in her shoes. I'm not pregnant," I say, "remember that."
>
> A social worker who trained me at Sloan-Kettering taught me something important: to clear my own agenda before I walk into the room, to let the patient set the agenda. It's the hardest lesson, and the most important one.

Despite this consciousness, counselors stand in a double-bind position to their clients. On the one hand, they are always making decisions about what sort and how much information a pregnant woman needs and can use and the form in which she can best absorb it. On the other, most of the information that the woman receives comes directly from the counselor, as she is unlikely to have a "folk model" of most of the diseases and risks associated with amniocentesis.[3] This is not true of any other aspect of pregnancy, or pregnancy loss, for which the process of medicalization is often contradicted by ideas and images shared in communities of women. The genetic counselor's communication about the health or illness of a potential child is shaped in a vocabulary that is exclusively medical, a grammar that is technological, and a syntax that has yet to be negotiated. Like many other workers in modern biomedicine, the genetic counselor really is the gatekeeper between science and social experience, regulating both the quantity and quality of the information on which decisions will be made.

At first glance, then, genetic counselors appear to control and shape communications because they hold almost exclusive access to the medical information on which its rules are predicated. But patients are not silent partners in these encounters, no matter how few sentences they utter. Ten percent of private patients and 50 percent of clinic patients break their appointments for genetic counseling, and somewhere between 20 and 50 percent of those counseled decide not to have amniocentesis.[4] Although reasons vary, they certainly include a disbelief or nonacceptance of the medical premises behind testing for fetal disabilities. This disbelief often takes the form of a clash of cultural assumptions, in which the pregnant woman says, "God will protect my baby," and the counselor says, "Most babies are born healthy, but 2 to 3 percent of all babies are born with birth defects." Sometimes, an interaction will reveal an especially clear instance of the shifting meaning of motherhood, with technocratic and traditional images uncomfortably located in the same sentence. As a low-income Chinese-American woman said, after her husband finished signing her up for the test and I queried her own desires, "My mother, my grandmother, they all had babies in China, and nobody did this. They wouldn't do it now, if they were here. Now is modern times, everyone wants to know everything, to know as soon as possible, in advance, about everything. What kind of information is this? I don't know, but I will soon have it, faster than I can understand it."

Most counselors insist that "I'm not here to sell amnios"; "I don't feel like a success or failure according to whether or not she takes the test." Their interests lie in informed consent, that particularly American legal-medical document that attests to a person's acceptance of information properly produced. Such an individual contract model is highly appropriate to a litigious society without a national health plan, where the only remedy for lack of information or services, or harmful information or services, is the malpractice suit. When viewed in this larger context, the counselor is providing protection for the doctor and the medical center, insuring their invulnerability despite the chaotic conditions of an "information revolution" in which the techniques and interpretations of genetic diagnoses are continuously under negotiation.

From a cultural point of view, the process of obtaining informed consent might represent a less legalistic and more communitarian project. Decisions concerning amniocentesis are made with the sum total of the knowledge a pregnant woman (and often her shadow network of support) brings to her interaction with a new medical technology. Culturally informed consent (or its refusal) is not reducible to the exchange of information-for-signature negotiated at the intake interview, for it is based on all the assumptions, fears, and norms concerning healthy and sickly children with which any given woman undertakes a pregnancy. It includes the meaning of illness in family history; the shame and pride attached to the bearing (or nonbearing) of children; beliefs about fertility, abortion, feminity, and masculinity; and the social consequences and prejudices surrounding disability, including the "courtesy stigmas" borne by those close to disabled people (Goffman 1963).

In this larger context, knowledge and power are not reducible to medical terminology, despite medicine's hegemonic claims. For surely this new technology has potentials that are at once emancipatory and socially controlling, depending on the context in which its use is shaped and practiced. Genetic counselors, no less than their various patients, are heirs to a eugenic script in which our aspirations for the liberation of women and children necessarily confront the current conditions under which family life is enacted. Only if the discourse on disability and reproductive rights is lifted out of the medical context and negotiated as part of popular culture, will it become possible to speak in another language.

We may occasionally catch glimpses of the effects of such displacements of medical culture in examining amniocentesis and disability among children as they are inscribed by mass media. To my question, "Where did you first learn about amniocentesis?" many women without advanced formal education answered "Dallas," "St. Elsewhere," or the *National Enquirer*. An Ecuadorian domestic worker tole me she'd learned about spina bifida (for which amniocentesis is offered) from the Jerry Lewis telethon. Several clinic patients gave articulate, up-to-date descriptions of children with Down's syndrome after Phil Donahue devoted a show to them. Middle-class amniocentesis patients often arrived armed with new questions after genetics stories appeared in the science section of the *New York Times*. As teaching and learning about inherited disabilities, and even amniocentesis, increasingly permeates the world of mass culture, medico-scientific discourse will have to confront its own popularization and challenges to it.

More widespread consciousness about both birth defects and technologies aimed at their screening may benefit both pregnant women and counselors. Pregnant women might then come to a decision from a more knowledgeable position, whereas counselors would be relieved of some of the burdens of protecting the medical hierarchies within which they now practice, in favor of more mutually constituted conversations with their patients. Under such circumstances, perhaps the shared responsibilities that surround education and decision making at the intersection of disability and reproductive rights could be discussed.

Acknowledgments

This chapter is drawn from an ongoing study, portions of which have been funded by the National Science Foundation, the National Endowment for the Humanities, and the Rockefeller Foundation's Program in Changing Gender Roles. I am deeply grateful to all of them. I especially thank the hundreds of women who shared their amniocentesis stories with me and the many health professionals whose commitment to better understanding their patients' experiences led them to cooperate in my research. All names have been changed to protect confidentiality. Helpful suggestions and criticisms were offered on earlier drafts of this essay by Robin Blatt, Robert Hahn, Alan Harwood, Marthe Gautier, Shirley Lindebaum, and Judith Stacey and by an anonymous reviewer for *Medical Anthropology Quarterly*. I thank them for their help in improving this work. Its weaknesses and conclusions remain my own.

Notes

1. The first NSCG workshop on "counseling the culturally different" was conducted by a team of minority genetic counselors at the annual meeting in 1985. Members of that team estimated that no more than thirty minority counselors are trained and practicing nationally. But definitions of both "minority" and "practicing" are open to interpretation.

2. Rothman (1986) argues forcefully that all prenatal diagnoses seem ambiguous to pregnant women.

3. Folk models for children with Down's syndrome—highly stereotyped and badly out-of-date—surely exist for most people. And some Black-Americans have opinions and images about the consequences of sickle-cell anemia. But virtually no one I have observed or interviewed knew about neural tube defects (e.g., spina bifida) for which the test is also done or about sex chromosome anomalies. With the exception of one Black-American woman who claimed the authority of dreams to accurately predict the birth of her disabled Down's Syndrome son and one white middle-class woman who knew that her third pregnancy would produce a disabled child because "it didn't move right, it didn't really move at all," no one who had the test, including women who subsequently received positive diagnoses of fetal disability, ever felt competent to predict the health status of the fetus she was carrying.

4. These numbers are drawn from the counselors and medical centers in which I did direct observations, and they vary enormously from facility to facility. A small percentage of "no-shows" have miscarriages before they are scheduled to speak with the counselor, but most have opted out of the appointment. In some clinics, patients

feel well served and can communicate directly with a nurse or paramedic about their desire to see, or not see, the genetic counselor. In others, suspicion, anger, or resignation dominate the experience of medical care, and appointments are coercively assigned and often broken in resistance. Those who choose not to have the test after counseling include women who discover that it can't diagnose the problem they came to discuss, as well as those deemed "appropriate candidates" who don't want it. Their reasons may include religious and spiritual beliefs; conflict with other family or community members about the meaning of pregnancy, testing, and having babies; and fear of the test, especially of needles (Hispanic women) and possible miscarriage (everyone).

References

Brewster, Arlene. 1984. After Office Hours: A Patient's Reactions to Amniocentesis. Obstetrics and Gynecology 64:443–444.

Brown, JoAnne. 1986. Professional Language: Words That Succeed. Radical History Review 34:33–52.

Goffman, Erving. 1963. Stigma: Notes on the Management of Spoiled Identity. Engelwood Cliffs, New Jersey: Prentice-Hall.

Kevles, Daniel. 1985. In the Name of Eugenics. New York: Knopf.

Reed, Sheldon. 1974. A Short History of Genetic Counseling. Social Biology 21:332–339.

Rollnick, Beverly. 1984. The National Society of Genetic Counselors: An Historical Perspective. Birth Defects 20:3–7.

Rothman, Barbara Katz. 1986. The Tentative Pregnancy. New York: Norton.

3

Ethical Issues in the Delivery of Quality Care to Pregnant Indigent Women

Marilyn L. Poland

Recent developments in reproductive technology allow early diagnosis and treatment of some maternal, fetal, and neonatal conditions including maternal hypertension, fetal anomalies, and problems associated with premature delivery. Complementary to early diagnosis and treatment is a system of health care that refers high-risk mothers to perinatal centers. These specialized centers provide genetic testing, prenatal diagnosis, ultrasonography, and other fetal tests. Studies of women who receive care and who deliver at high-risk perinatal centers indicate that these women have healthier babies compared with those of comparable risk delivering at a community hospital without the benefit of technology (Institute of Medicine 1985).

Despite the development of this complex and expensive system of medical care, high rates for infant mortality and low birthweight continue; moreover, such high rates are disproportionately common among black and low-income Americans. Reasons for these rates are unclear, although social-class differences affecting health habits and access to health care are thought to play a major role. For example, poor women may not have the resources to seek preventive health care early in pregnancy, when some health problems can be detected and treated. Use of preventive health care among the poor has been characterized as generally episodic and without a consistent health-care provider (Rundall 1979). In addition, lack of an adequate social support system may hamper an individual's ability to seek health care. Boone (1982) studied poor, black mothers who delivered low-birthweight infants and found that these women were more likely to feel isolated from family, friends, and public agencies and that many received little prenatal care. The Michigan Department of Public Health (1984) reported that 9 percent or more of mothers in Detroit received little or no prenatal care, although care was available at a facility near their homes. These mothers were also often at great risk of developing complications of pregnancy and may have benefited from early access to prenatal care and fetal technology.

Two surveys were undertaken to examine reasons why women do not receive adequate prenatal care and to explore the ethical issues relating to access to prenatal care that are raised by these studies. In one survey, reasons were assessed as to why women received varying amounts of prenatal care. In a second survey, the perceptions of women were compared to the perceptions of obstetricians as to what constitutes medical risk during pregnancy.

A Study of Prenatal Care Among Poor Women

One hundred and eleven women who delivered at Hutzel Hospital in Detroit during 1985 were interviewed two to five days postpartum, and their medical charts were examined for sociodemographic and medical factors (Poland, Ager, and Olson 1987). These women were all poor and most (88 percent) were black. They varied considerably in age (sixteen to twenty-four years), number of children (one to seven), number of prenatal visits (none to twenty-one), and in pregnancy outcomes. Although none of the babies died, 25 percent weighed less than 5½ pounds. These babies are labeled as "low birthweight," a category associated with an increased chance of death and developmental disabilities. Fifty-seven of these women were listed in hospital records as "walk-ins"—patients who had not received prenatal care by a physician associated with the hospital—and fifty-four were women who had attended the hospital's prenatal clinic. Patients listed as walk-ins were selected at random from hospital charts; a comparison case of a woman receiving prenatal care at the hospital was selected in a similar fashion.

For purposes of analysis, these patients were further categorized according to the amount of prenatal care they received, based on a formula that takes into account the month of gestation in which they registered, number of times they were seen, and the length of gestation (Kessner 1973). Thirty-five women who had adequate care registered early and kept their appointments. Women with intermediate amounts of care either registered late or missed several appointments. The forty-one defined as receiving inadequate care had little or no prenatal care. The interview consisted of questions that were both open ended and fixed choice to assess the experience of pregnancy and prenatal care over the current and previous pregnancies. The interview included: descriptions of how and when a woman knew she was pregnant; who she talked to about the pregnancy, advice she received; names, dates, number of appointments, and descriptions of all prenatal-care sources; problems in receiving care; and descriptions of how and why she decided to seek or not seek care. The content of responses were analyzed, and they were summarized by amount of prenatal care received. Pregnancy information from the medical chart was also coded for analysis.

The data were analyzed to compare: 1) demographic and medical factors by care group; 2) sociocultural factors by care group; and 3) beliefs about what constitutes "risk" held by the women versus criteria commonly used by obstetricians. For the first analysis, several significant differences between the three groups were found for several demographic and health factors: age, parity, risk of complications, length of pregnancy, and birthweight. Women who received inadequate care were older, had more children, were at high risk for complications, experienced shorter pregnancies, and produced smaller babies than women in the other two groups.

There were also sociocultural differences between care groups. First, women who received little or no prenatal care were less interested in their pregnancies, were more easily frustrated by barriers to care such as lack of transportation, and felt a high degree of isolation from friends and health professionals. Many contemplated abortion but were either advised against it, did not have the funds to pay for an abortion, or were too late in gestation to have an abortion. Some delayed making the decision to abort until it was too late or, as one woman said, "it was easier to have it." In addition, many of these women who qualified for Medicaid did not seek it because they did not want to bother with "hassles" involving visiting the social worker and filling out complicated forms. Finally, women in the inadequate-care group delayed significantly longer in telling others about their pregnancies than did the other women in the study and received more negative responses. This delay correlated with comments the women made in which they described having few friends or relatives to help them and a general feeling of isolation. These women were also more likely to come alone to the hospital when they were in labor. Three came on a bus because they could not afford a cab and did not have anyone to drive them there.

The second sociocultural difference between groups was the lack of dependence on information and advice about pregnancy from professionals and from written materials. For example, although the signs by which women knew they were pregnant did not differ, women in the inadequate-care group suspected they were pregnant significantly later than those in the other two groups. They were less aware of body changes and what these meant. Many women used family, friends, and professionals as sources of information about pregnancy, but women in the inadequate-care group depended more often on knowledge gained during previous pregnancies and less often on written information. In fact, several women in the inadequate-care group registered late for care because they could not read well enough to fill out insurance forms and were intimidated by registration procedures, which included a twenty-one page application form for Medicaid.

A third difference between groups related to the facility at which the women sought prenatal care. Although women in the adequate-care and intermediate-care groups were more likely to use more than one source, 30 percent of these women and almost all of those receiving inadequate care used a twenty-four-hour emergency drop-in center for some or all of their prenatal care. These centers were originally established for short-term, minor problems. In Detroit, they are used by almost 40 percent of pregnant women who use public agencies for health care (Waller 1986). Although many of the women felt that these centers did not provide high-quality prenatal care, they used them because they were close to home, were inexpensive, were accessible without an appointment (some women did not have a phone), and were always open. In addition, the women could be "checked by a doctor" when they felt the need, and some centers provided prenatal vitamins at less cost than drugstores did. But their major attraction was their location in poor neighborhoods. Most (65 percent) of the women in this sample did not have access to a car. Detroit buses are their major source of transportation, and bus routes define the boundaries of available stores and services. The buses are often unreliable, located a distance from a woman's home, require one or more transfers, and are viewed as unsafe for pregnant women to use in bad weather or when buses have dangerous passengers. Thus, drop-in centers represented an easy alternative to the problem of city transportation.

According to the women, the quality of medical care at drop-in centers was inferior to that offered at prenatal clinics. They reported that blood-pressure measurement and blood tests for anemia were the major procedures used to monitor their pregnancies. Although these procedures are part of routine prenatal care, there are many other methods routinely employed to assess the progess of pregnancy, especially in women who are poor and at higher risk for abnormalities. Most of the women who sought care at emergency drop-in centers were not monitored for problems other than hypertension and anemia.

The fourth area of difference between groups was the women's perceptions of the value of prenatal care and their attitudes toward doctors. Women in the inadequate-care group varied in their responses about the importance of prenatal care, but, in general, they valued prenatal care less than those in the other groups. Many women believed that if they felt all right, the baby was moving, this was not a first pregnancy, and they took their prenatal vitamins, they did not have to see the doctor unless there was a problem. As one woman said, "Poor people can't afford to see the doctor all the time when nothing is wrong." Some women, however, did not share this view. As one said, "I was scared the whole time. If I could have seen a doctor, it would have eased my mind."

Women who had less care also had more negative attitudes toward doctors. Women in all three groups described having to relate to doctors, nurses, and others who were "nasty," "had bad attitudes," "cursed at them," or "treated them poorly" because they had Medicaid or no insurance. Seeing a different doctor each time, as most women did, did not foster a positive relationship. Women were also critical of doctors who did not answer their questions or treated them with disrespect. One woman said, "Pregnancy is a scary time. Pregnant women should be given special attention and not treated like cattle."

Written notes by doctors and nurses about women who received little care often contained derogatory phrases and corroborated this negative view. Women who received inadequate care were labeled as such in the chart, and those with no care, or care received at drop-in centers, were designated as walk-ins on both the outside of the medical chart and as part of the identifying information at the top of each page. These women were referred routinely for social service review to assess the home conditions and the mother's ability to care for her infant. One woman who was mislabeled as a drop-in was resentful of that label and quick to point out to the interviewer that she received care from her family physician and was not irresponsible "like some of those other women."

A subset of forty-one mothers from our sample was assessed for their perceptions about what constitutes medical risk during pregnancy. These perceptions were compared with the perceptions of obstetricians on a high-risk perinatal scoring system that is used in many perinatal centers (Hobel 1973). As is shown in Table 3.1, there were many discrepancies between what women and obstetricians perceived as risk. For example, the women felt that not taking prenatal vitamins represented a major health risk, whereas having five or more babies or having a previous baby weighing less than 5½ pounds did not represent risk. These opinions are in direct opposition to the views of the medical system upon which referral to perinatal centers is based. Thus, differences in perceptions of what constitutes risk of complications of pregnancy may also contribute to why some women seek less prenatal care or seek care at a convenient drop-in center.

These two surveys found a variety of reasons why women failed to receive adequate prenatal care, and they also found an association between receiving inadequate care and having a low-birthweight infant. In addition, the results suggested that those with the highest risk of complications may receive the least amount of prenatal care. Personal barriers to receiving prenatal care included: failure to seek insurance even though most of these women were eligible for Medicaid; health beliefs and attitudes such as not wanting the pregnancy, fear of doctors, and assigning a lower value to prenatal care; and beliefs about risk that were often at variance with standard medical practice. There were also significant differences in the amount of

TABLE 3.1

Perceptions of Risk by Mothers and Physicians

Risk Category	Physicians' Perception[a]	Mothers' Perception[b]
High risk	High blood pressure Had a baby weighing 5½ lbs	High blood pressure Anemia (low blood) Not taking prenatal vitamins
Medium risk	Had 5 or more babies Over 35 years old Being obese (200 lbs) Flu in early pregnancy Anemia (low blood)	Flu in early pregnancy Had one previous abortion Over 35 years old Being obese (200 lbs)
Low risk	Had one previous abortion Gaining over 20 lbs Not taking prenatal vitamins	Having 5 or more babies Had a baby weighing 5½ lbs Gaining over 20 lbs

a. Based on POPRAS high-risk scoring system (Hobel 1973).

b. Ranges for all factors varied from high to low.

emotional and tangible support women received; those in the inadequate-care group felt isolated from a network of friends and relatives as well as health professionals.

Ethical Issues Raised by the Study

The surveys identified barriers to the utilization of medical technology by women who may be most likely to benefit from its use. The findings raise two ethical issues. First, how accessible should prenatal-care technology be? In this question, accessibility implies both the availability of this technology and the special social supports necessary to encourage women to receive it. The second issue pertains to the general social norm in this country that prenatal care is important and necessary for the health of the fetus and that mothers should be responsible for obtaining it. In view of this social norm, will women retain the right to refuse prenatal technology when their values and beliefs differ from those of medical experts who urge its use?

Accessibility of Prenatal-Care Technology

The United States is unsurpassed in its development of perinatal technology to promote healthier babies. However, although considerable effort and resources have been expended to develop and test new technology, there has been less interest in ensuring the availability of even routine prenatal care to all women in need. Despite an improved economic climate, poor women actually received less prenatal care in the past five years than previously (Children's Defense Fund 1984). In addition, the infant mortality rate continues to be twice as high for blacks than it is for whites in the United States (Kleinman 1985).

Many of the women who participated in our first survey went to considerable effort to seek prenatal care. Most had to overcome obstacles such as the difficulties of using public transportation to reach overcrowded clinics where they had to wait for long periods before being treated. Often they were treated by a different doctor at each visit. Those who lacked the personal resources to seek prenatal care at specialized perinatal centers sought care at emergency drop-in clinics that was inferior to the care offered by prenatal clinics or received no care at all.

The notion of providing quality prenatal care to all women as a basic health right is an ethical concept indicating that society places a high value on human life. The emphasis in the United States is on developing expensive technology to diagnose or to treat fetal newborn conditions. There is less emphasis on developing the social mechanisms to encourage women to use the technology or to seek preventive health services that might reduce the need for expensive neonatal intensive care.

A 1985 survey of twenty-three countries in the European region of the World Health Organization described various forms of maternity support and protection as national policy (Wagner 1985). These supports included health insurance; paid maternity leaves of at least twelve weeks; child-care allowances; special privileges when traveling by public transportation; priorities for loans and housing; permission to change jobs; free milk, vitamins, baby equipment; special working hours; and numerous other benefits. None of these supports is routinely available in the United States. Although the United States leads the world in the development of perinatal technology, there are no national policies that insure its accessibility. All twenty-three developed countries have a gross national product far below that of the United States, but all have made a commitment to maternity protection.

The infant mortality rate in the United States was 10.9 infant deaths per 1,000 live births in 1984, which compares unfavorably to a rate of 7 in Sweden and a rate of 8.1 in Japan (Hartford 1985). In these countries, with less-developed perinatal technology, all women receive prenatal care. That these countries provide this basic health service reflects national concern and responsibility for the welfare of pregnant women and plays a major role in maintaining low infant mortality rates.

Valuing of Prenatal Technology

The second ethical issue centers around problems arising from differences in perceptions of the value of prenatal technology. The general social norm is that prenatal care is necessary for fetal health; therefore, any woman who does not seek care is jeopardizing her fetus and acting irresponsibly. Our study reported that women who received inadequate prenatal care did so for a variety of reasons, such as a sense of social isolation, rejection of the pregnancy, and fear of doctors and procedures. Some health professionals expressed their anger and frustration with these women in written notes placed in the medical charts. In Chapter 1 of this book, Jordan and Irwin describe cases in which women who refused surgery were ordered to undergo cesarean sections by medical and legal authorities. As the efficacy of receiving prenatal care for a healthy baby becomes unquestioned, a similar fate could await women who fail to seek prenatal care.

As technology makes the fetus more visible, more real, and more like a patient, health professionals and others feel an increased responsibility for its welfare. Some maternal behaviors, such as drinking alcohol or using drugs during pregnancy, are being viewed with alarm as technology demonstrates the harmful side effects of these behaviors (Sokol 1981). Seeking prenatal care has also been viewed as the norm and as a maternal responsibility (Shaw 1984). How will society see its responsibility to pregnant women who do not share middle-class values, life-styles, or resources and do not seek prenatal care? Will social sanctions be placed against pregnant women who are deemed irresponsible, such as those who do not seek prenatal care early and often?

References

Boone, Margaret. 1982. A Socio-Medical Study of Infant Mortality Among Disadvantaged Blacks. Human Organization 41:227–236.

Children's Defense Fund. 1984. American Children in Poverty. Washington, D.C.

Hartford, Robert. 1985. Comparative Overview of Trends and Levels in Procedures of the International Collaborative Effort on Perinatal and Infant Mortality. Washington, D.C.: National Center for Health Statistics, pp. 89–95.

Hobel, Calvin, M. A. Hyvarinen, D. M. Okada, and W. Oh. 1973. Prenatal and Intrapartum High Risk Screening: Prediction of the High Risk Neonate. American Journal of Obstetrics and Gynecology 117:1–9.

Institute of Medicine. 1985. Preventing Low Birth Weight. Washington, D.C.: National Academy Press.

Jordan, Brigitte, and Susan L. Irwin. 1988. "The Ultimate Failure: Court-Ordered Cesarean Section." Chapter 1 in this volume.

Kessner, A. M., A. Auiger, C. E. Kalk, and E. R. Schlesinger. 1973. Infant Death: An Analysis by Maternal Risk and Health Care. Washington, D.C.: Academy of Sciences, Institute of Medicine.

Kleinman, Joel. 1985. Perinatal and Infant Mortality. Recent Trends in the United States in Procedures of the International Collaborative Effort on Perinatal and

Infant Mortality. Washington, D.C.: National Center for Health Statistics, pp. 37–55.

Michigan Department of Public Health. 1984. Vital Statistics 1982. Lansing, Michigan.

Poland, Marilyn, Joel Ager, and Jane Olson. 1987. Barriers to Receiving Adequate Prenatal Care. American Journal of Obstetrics and Gynecology 157:297–303.

Rundall, Thomas, and John Wheeler. 1979. The Effect of Income on Use of Preventive Care: An Evaluation of Alternative Explanations. Journal of Health and Social Behavior 20:397–406.

Shaw, Margery. 1984. Conditional Prospective Rights of the Fetus. Journal of Legal Medicine 5:63–116.

Sokol, Robert. 1981. Alcohol and Abnormal Outcomes of Pregnancy. Journal of the Canadian Medical Association 125:143–148.

Wagner, Marsden. 1985. Having a Baby in Europe: Lessons for North America. Paper presented at 113th Annual American Public Health Association, Washington, D.C.

Waller, John, Steven Blount, Paul Giblin, Marilyn Poland, and Alan Reed. 1987. An Analysis of Births, Infant Morbidity, and Infant Deaths as Outcomes of Prenatal Care Delivered by Inner City Detroit Clinics. Paper presented at 114th Annual American Public Health Association, Las Vegas, Nevada.

PART TWO

Ethical Decisions in the Treatment of Newborns

4

Life-and-Death Decisions and the Sociology of Knowledge: The Case of Neonatal Intensive Care

Renee R. Anspach

The Problem

Entering an intensive-care unit for newborn infants for the first time can provide the visitor with what may appear to be a surrealistic encounter with the twenty-first century. The diffuse din of the monitors, the many incubators and intravenous lines, the eerie glow of the ultraviolet lights used to combat jaundice—the elaborate machinery of the nursery—provide a stark contrast to the tiny patients. Next to each incubator stands a nurse who from time to time may reach into a porthole of the incubator to administer medications or to take vital signs. At least twice a day, an entourage of physicians stops by each incubator, making daily rounds. The newborn intensive-care unit stands as a monument to science and technology, a living testament to the vast resources that our society has committed to saving life at its beginning. There are, however, times in the social life of the intensive-care nursery when neonatal intensive care may seem less a symbol of progress than a confrontation with the problematic. These moments occur when parents and health professionals, faced with an infant who is terminally ill or who may survive with serious handicaps, must decide whether the infant will live or die.

In this chapter, I consider decisions in which the central issue is the prognosis of an infant whose life is in question. Perhaps the most difficult and controversial of life-and-death decisions, such decisions raise a dilemma emblematic of neonatal intensive care as an enterprise: what to do in the absence of sound information concerning the future of the infant whose fate is being decided. Basing my argument on data collected in the course of sixteen months of fieldwork in two intensive-care nurseries, I will suggest that life-and-death decisions may be approached from the standpoint of the sociology of knowledge, a perspective that relates ideas to the social context in which they occur. Three related points will be emphasized. First, I will show how the decisions of health professionals are shaped by the practical circumstances of their work.[1] I will present data, taken from interviews and case studies,[2] that suggest that members of the nursery staff, because of their differing work experiences, may arrive at conflicting conclusions about the prognoses of infants whose lives are in question; these differences frequently result in conflicts concerning how life-and-death decisions should be made. Second, I will show how the organization of the intensive-care nursery as a work environment structures the perspectives of those who work within it. Third, in a more speculative vein, I will suggest that, contrary to what is commonly believed, much of what appears to be "wrong with," or problematic, about life-and-death decisions is not the "fault" of the individuals who make them but is instead located in the very way in which the intensive-care nursery is organized. This chapter, then, is concerned with the relationship between the knowledge involved in prognosticating and reaching life-and-death decisions and the social organization of the newborn intensive-care unit.

Theoretical Considerations

The perspective used in this study differs from two approaches to life-and-death decisions: attempts by bioethicists to elucidate ethical issues and sociological studies that identify the norms and values at stake in life-and-death decisions.

Much of the literature about life-and-death decisions has been written from a bioethical perspective. (For a discussion of this approach to decisions in neonatal intensive care, see Chapter 7.) Bioethics does not examine data that describe the way in which decisions *are* made but rather is concerned with the way in which decisions *should* be made—and is, thus, a disciplined incursion into the realm of the elusive "ought."[3] Writers from this perspective have generally assumed that decisions reside in the individual, who reaches them alone, apart from institutional constraints. Presumably, optimal decision making would ensue if individuals would become more reflective in their application of ethical principles. This assumption lends bioethics a reformist quality that seeks solutions in disciplined reflection rather than in changes in the social milieu.

In contrast to bioethical analysts, sociologists examine the way in which decisions actually are made rather than the way in which they should be made. Sociologists who have studied neonatal intensive care have approached life-and-death decisions from a normative standpoint (Wilson 1970). This view, written largely from a structural-functional perspective, assumes that life-and-death decisions are the product of consensual norms and values, transmitted through professional socialization. These sociologists note that the traditional norm of aggressive intervention (Parsons 1951) on the part of the physician is undergoing reexamination, and they posit cultural reasons for this transformation. For example, it has been suggested that medicine is moving from an ethic based on unconditional sanctity of life to one premised on quality of life (Fox and Swazey 1973; Fox 1974). Crane's (1975) major study of physicians' decisions to treat critically ill infants is an attempt to identify new normative patterns that have emerged. In this study, an attitude survey examining how physicians respond to hypothetical dilemmas (posed in a questionnaire), Crane does not observe decisions directly but does discuss decisions concerning newborn infants. Crane concludes that the major criterion used by physicians in reaching life-and-death decisions is the patient's social potential or capacity to perform social roles.

In sum, life-and-death decisions have been examined from two perspectives. The first view locates decisions in the individual conscience; the second locates them in the collective conscience. Both perspectives share a common set of limitations. First, both focus on decisions in which the infant's prognosis is known or predictable. For example, congenital anomalies such as Down's syndrome, spina bifida, Tay-Sachs disease (a serious genetic disease in which those affected die in infancy), and anencephaly (born without a brain) have been discussed frequently, although many newborns have a prognosis that is uncertain (Jonsen and Lister 1978). Thus, although bioethicists and sociologists have emphasized conditions that are ethically complex and prognostically simple, actual decisions are often complex from the standpoint of both prognosis and ethics. Second, both perspectives focus their analysis on a single decision-maker—usually the physician; less often, the nurse. This implies that the actor reaches decisions alone apart from the influences of others. Finally, both approaches present a rather idealized image of decision making and emphasize the principles in life-and-death decisions rather than the process by which they are reached or the social context in which they take place. Decisions, however, are complex social phenomena that take place amid interactions, organizations, institutions, and power relations.

My approach to life-and-death decisions from the standpoint of the sociology of knowledge derives its guiding imagery from the sociology of occupations and organizations. As Mannheim (1937) suggested, people do not have an "objective," all-embracing view of reality but rather derive their perceptions from the social circumstances in which they find themselves. The social milieu not only structures how participants construe the "facts" to be used in reaching life-and-death decisions but determines what facts are viewed as salient and what is designated as data. Thus, the social milieu of the newborn intensive-care unit shapes the decisions of the various occupational groups that work within it.

In what follows, I will examine those decisions in which there was conflict concerning the prognosis of the infant whose life was in question. In order to make any life-and-death decision, the participants must have some idea about the kind of future that awaits the patient whose fate is being decided. However, owing to the highly innovative, "experimental" nature of neonatal intensive care, these judgments often prove notoriously difficult. For example, in some nurseries it is possible to resuscitate infants as small as 600 grams (1.1 pounds) and as young as twenty-five weeks gestational age, but many of these infants die, and the ones that do live risk damage to major organ systems, including serious brain damage. Thus, it often becomes difficult to predict the viability or intellectual potential of many newborn infants. Although many nurseries have undertaken follow-up studies to evaluate the degree of morbidity that occurs in their patients, these studies do little to clarify matters for those facing immediate life-and-death decisions (Jonsen and Lister 1978).

In both intensive-care nurseries that were studied, a broad consensus around a principle emerged: If practitioners were reasonably convinced that an infant was unable to survive or would survive with serious neurological defects (such as severe mental retardation or cerebral palsy), then it was considered appropriate to withhold life-sustaining treatment. However, even when participants could agree on the *principle* involved in a life-and-death decision, all too often they could not agree on the *prognosis* of the infant whose life was in question.

When the underlying dilemmas were prognostic, participants merely acknowledged difficulties in making predictions. More often than not, however, the prognoses of infants whose lives were in question became the subject of debate, disagreement, and even acrimonious conflict among attending physicians, residents, and nurses. These conflicts followed recurrent patterns: Most often it was the nurses—less frequently, the residents—who were the first to conclude that the infant would not recover or would have severe mental disabilities. This pattern suggested the hypothesis that, in situations in which prognostic ambiguity is caused by a lack of precise "scientific" knowledge, nurses, residents, and attending physicians may have different methods for arriving at prognostic assessments. Each group of participants may, in fact, be approaching the life-and-death decision from the standpoint of a very different data base. The fact that conflicts surrounding prognosis seem to have a systematic and patterned nature alerts us to the fact that predicting and prognosticating, an ostensibly scientific activity, may in fact have a social and organizational basis.

In the following pages, I will discuss the way in which the newborn intensive-care unit qua organization allocates different types of information to those who work within it, and, in so doing, defines the character of the data that each group brings to the life-and-death decision. Each occupational group has a different set of daily experiences that define the contours of the information used in making prognostic judgments and in reaching life-and-death decisions. First, the attending physicians (neonatologists) are academicians as well as clinicians, and they alternate their rotations with research responsibilities. Although attending physicians spend the least absolute time in the nursery observing individual patients, they can draw upon a reservoir of clinical experience and have unique access to information from the follow-up study. Second, the residents and fellows, during their month-long rotations, spend considerable time in the nursery. However, their contact with infants is limited, episodic, and confined to technical interventions. Last, the nurses, although they are subordinate in the occupational hierarchy, nevertheless play a significant role in life-and-death decisions. Unlike the other participants, the nurses sustain continuous, close, and long-term contact with infants who are their patients. Moreover, they engage in types of patient care (such as infant stimulation) that transcend the technical.

Given these very different experiences, one would expect to find each occupational group construing the facts in different ways and bringing differing notions of what constitutes adequate data, or evidence, to the decision-making process. I am suggesting, then, that the organization serves as a sort of interpretive lens through which its members perceive their patients and predict their futures and, therefore, functions as an *ecology of knowledge*.

The Organizational Basis of Prognostic Judgment

A number of sources of information may be used in making predictions in medical settings; each differs in the degree of patient contact that is required. *Technological* cues refer to information obtained by means of diagnostic technology. *Perceptual* cues refer to information gathered through direct perception of the patient, including palpation, percussion and, most commonly, observation. *Interactive* cues refer to information arising from the social interaction between patient and practitioner. In adult medicine, the major mode of interaction is the clinical history. However, if one defines social interaction broadly to include nonverbal aspects of communication, it is evident that social interaction can take place between infants and practitioners.

In order to explore their methods of prognostication, residents and nurses were asked, "How can you tell if an infant is doing well or poorly?" Responses given in an interview situation obviously may not reflect actual practice, but taken together with field observations they do provide useful information. My hypothesis was that physicians, because of limited contact with patients, would be more likely to rely upon diagnostic technology, whereas nurses would be more likely to mention interactive cues in their responses.

Table 4.1 presents the frequencies with which technological, perceptual, and interactive cues were cited. All of the occupational groups relied on technological cues and on perceptual cues. This suggests that the value of diagnostic technology assumes a superior epistemological status in the intensive-care nursery. However, the most striking contrast concerned the frequency with which interactive cues were noted. A substantially larger number of nurses than physicians mentioned interactive cues in their responses. This difference, although statistically significant in the Randolph nursery, was not significant in the General nursery, as the data in Table 4.2 suggest.

In order to understand the nature of the prognostic process, it is necessary to go beyond gross tabulations and examine the interviews. Most residents and fellows based their diagnostic judgments on a combination of technological and perceptual cues. As would be expected, these technological cues included changes in weight, laboratory results, and measures of respiratory function. What is significant, however, is that these physicians cited only those perceptual cues that could be obtained by means of a cursory physical examination, which testifies to the limited nature of contact between physicians and patients. One resident lamented that the reliance on diagnostic technology was at odds with the ideals of his medical training:

> Well you really get dramatic kinds of information, I mean the baby's blood gases can improve or deteriorate markedly. I think that you can tell best in the nursery by your clinical exam unless something really gross supervenes, like congestive heart failure, and even more by how his laboratory parameters are doing . . . is the child gaining weight, are they having apneic spells [episodes in which the infant stops breathing], you know, do his electrolytes, does his c.b.c., do his blood gases look good, and in a way that's frustrating, because when you are a physician you like not to depend on the laboratory, but the bottom line, apart from information you can obtain from weights and things like that, *the bottom line is what are this kid's values?* [italics are mine].

Like the other participants, nurses cited technological and perceptual cues, but the major difference was the frequency with which they cited interactive cues. These included facial feedback, eye contact, smiling, and other aspects of the infant's facial expression. The theme of responsiveness was mentioned frequently by those nurses who discussed interactive cues, including responses to visual, tactile, and auditory stimuli:

> When an infant's breathing on his own, it becomes a little more subjective and some of the signs I use to judge are: level of awareness, how alert I

TABLE 4.1

Types of Information Used by Residents, Fellows, and Nurses in Patient Assessments

(RELIABILITY: (92%))

NUMBER OF RESPONDENTS NOTING THE FOLLOWING TYPES OF INFORMATION	A. RANDOLPH NURSERY OCCUPATION			B. GENERAL NURSERY OCCUPATION			C. BOTH NURSERIES OCCUPATION		
	Residents	Fellows	Nurses	Residents	Fellows	Nurses	Residents	Fellows	Nurses
Technological Cues[a]	12	3	16	7	3	7	19	6	23
Perceptual Cues[b]	9	2	15	5	4	7	14	6	22
Interactive Cues[c]	2	0	12	1	0	4	3	0	16
Other[d]	2	1	1	1	0	0	3	1	1
Total Responding	14	3	18	7	4	9	21	7	27

Explanation of Coding

The frequencies add up to more than the total number of respondents, since respondents typically cited more than one type of information.

a. Technological cues refer to any type of information obtained by the use of measurement instrument, including the most simple (e.g., a thermometer or scale). The technological cues that were cited most frequently include: laboratory data (electrolytes, bilirubin), bloodgases, weight gain, vital signs, respirator setting, monitors, and radiological evidence.

b. Perceptual cues include any information obtained by means of the clinician's perceptions of the patient. Most frequently noted were color, activity level, muscle tone, and "soft" neurological signs (posturing, deviation of eyes).

c. Interactive cues are those which emanate from the interaction between patient and practitioner, and which relate to the patient's ability interact socially. Most commonly cited interactive cues include: Facial expression (e.g., smiling, eye contact); responsiveness to tactile, and visual, and auditory stimuli.

d. Other types of information include asking the nurse's and the practitioner's impression of the family.

TABLE 4.2

Types of Information Used in Patient Assessments by Occupation

NUMBER OF RESPONDENTS WHO:	A. RANDOLPH NURSERY OCCUPATION			B. GENERAL NURSERY OCCUPATION			C. BOTH NURSERIES OCCUPATION		
	Physicians (Residents & Fellows)	Nurses	Total	Physicians (Residents & Fellows)	Nurses	Total	Physicians (Residents & Fellows)	Nurses	Total
Did not mention Interactive cues[a]	15	6	21	10	5	15	25	11	36
Mentioned Interactive cues[b]	2	12	14	1	4	5	3	16	19
Total	17	18	35	11	9	20	28	27	55
	$X^2 = 6.88$ df = 1 $P < .01$			Fisher's exact test[c] P = .097 n.s.			$X^2 = 12.259$ df = 1 $P < .001$		

Notes

a. These respondents cited technological cues, perceptual cues or "other" types of information only.

b. These respondents mentioned interactive cues in conjunction with technological cues, perceptual cues, and "other" types of information.

c. The Fisher's exact test was used because of the very small sample in the General Nursery. X^2 distributions, even when corrected for continuity, are likely to produce distortions when applied to very small samples (Blalock 1960:221-225).

> think he is, how he responds to noises in the room, to my touching him, to my knocking against the isolette, any kind of tactile or auditory stimulation, how active he is when he's left alone, whether he moves on his own, or whether he doesn't respond unless stimulated. Mentation, level of awareness means a lot to me.

Another frequently cited theme was the ability of the infant to respond to the attention of the nurses:

> I think if they're doing well they just respond to being human or being a baby. . . . [Can you give me examples of how babies interact?] Basically emotionally if you pick them up the baby should cuddle to you rather than being stiff and withdrawing. Do they quiet when held or do they continue to cry when you hold them? Do they lay in bed or cry continuously or do they quiet after they've been picked up and fed . . . do they have a normal sleep pattern? Do they just lay awake all the time really interacting with nothing or do they interact with toys you put out, the mobile or things like that, do they interact with the voice when you speak?

This respondent explicitly identified the procedures she uses to determine "appropriate" emotional responses from the baby. What appears to organize her interpretations is an implicit criterion for the attribution of humanhood: Infants should be able to interact in such a fashion that their actions may be seen as intentional. These interpretations, of course, transcend the conventional notion of signs, symptoms, or disease.

Nurses differed from other respondents in another way. They often characterized their knowledge as tacit, taken for granted, intuition:

> It's just a gut feeling. I think that with a lot of nurses that have been here for awhile, the gut feeling has developed from a lot of observation that they have recorded subconsciously and never put on paper, and that no one's ever written about and I think that those are a lot of those things, so when people say, "I don't like that kid, there's something wrong with him," the gut feeling's really based on a log of information study that they don't notice what they're doing.

Gut feelings signify a different legitimation of knowledge than what is designated as scientific evidence. To portray one's knowledge as a gut feeling is to introduce information that, on the one hand, is shielded from debate (it is not necessary to document or defend what is called a "gut feeling"), but, on the other hand, carries little weight within scientific argument and is excluded from the realm of scientific discourse. Gut feelings are, to paraphrase Pascal, "reasons of the heart." In short, the nurses relied upon cues gleaned from their interactions with infants, which they interpreted within an interactive frame of reference and characterized as "gut feeling."

There is, of course, an alternative explanation for these findings. It might be argued that the differences in the knowledge of physicians (residents and fellows) and nurses results from the prior sex role socialization of men and women. Because the studies included no male nurses, it is impossible to resolve this issue. However, as there were no differences in the responses of female and male residents, there is reason to accept the more parsimonious explanation: that the contrasting responses of physicians and nurses reflect their experiences in the organization rather than prior sex role socialization.

Like the nurses, attending neonatologists spoke of having gut feelings or clinical experience that would enable them to make predictions at a glance. However, these gut feelings were entirely perceptually based: "Of course I have gut feelings, based on clinical experience. I can often walk into a nursery and tell you how any patient might turn out, based on the patient's diagnosis and appearance. I've seen it so many times before. But I've also been around long enough to know I can be wrong." However, attending physicians, unlike nurses, placed little confidence in their intuitive judgments. Thus, on no occasion did I ever observe an attending physician reach a decision to withdraw life support until diagnostic tests had been performed. Even when attending physicians had strong negative intuitions about a baby's prognosis, they would continue to support the baby until tests had been conducted. Perhaps this action was related to the legal demand for documentation. There is one other way in which the attending physicians differed from the other participants: They had unique access to information, often unpublished, from the nursery's follow-up study. The fact that the attending physicians had a virtual monopoly on this type of information had the effect of increasing their authority in life-and-death decisions.

The Confrontation Between Modes of Knowing: An Illustrative Case

The following case illustrates the consequences that can ensue in the face of prognostic dilemmas. Robin Simpson's birth in May 1978 was believed to be the result of a domestic tragedy. During a quarrel with her husband, Mrs. Simpson fell down the stairs; nine days later, she entered a small private hospital, showing signs of impending premature labor. Five days after that, Robin—who weighed 2 pounds, 5 ounces—was delivered nine weeks prematurely by cesarean section. Months later, physicians and nurses in the Randolph nursery were to recall these unfortunate events in an effort to make sense of the unusual course of an illness that kept Robin in the intensive-care nursery for seven months. During this time, Robin was never able to exist apart from the life-support equipment necessary to sustain her breathing. Robin's protracted stay in the nursery was also costly, culminating in hospital bills of a quarter of a million dollars.

But when Robin first arrived in the nursery, she appeared to be an infant who was doing reasonably well. However, beginning one day after her birth, it became increasingly apparent that Robin's case was unusual, as she repeatedly violated expectations of a "normal course" for a premature infant with lung disease. She developed lung disease much later than would be expected and began to deteriorate rapidly. Moreover, in the days, weeks, and months that followed, Robin was unable to breathe without the total support of the respirator, and attempts to wean her met with repeated failure. Faced with Robin's total lack of progress, physicians and nurses began to wonder whether her lung disease was irreversible or uncurable. By August 1978, when Robin had been dependent on the respirator for three months, a conference was held in which the staff made the difficult decision to wean her from the respirator, leaving her in an oxygen tent to support her breathing. The parents reportedly faced the news of this decision to allow Robin to "sink or swim" with a certain amount of relief. No one expected Robin to survive without the support of the respirator. But she began to breathe on her own, again defying expectations.

Three more months elapsed. Robin's lung disease showed no signs of improvement. The nurses had become increasingly pessimistic about her chances for recovery and concerned about her mental development. Frustrated with the prospects of caring for Robin, they began to withdraw. There were other problems as well. The Simpsons' marriage was reported to be deteriorating. Mrs. Simpson, who at one time had visited regularly, was seen only rarely in the nursery. In November 1978, after Robin had been in the nursery for seven months, the head nurse discussed Robin's case with Dr. Nelson, the attending neonatologist, and, once again, a conference was convened to decide Robin's fate.

The story of Robin Simpson's stay in the nursery raises social and ethical issues encountered in intensive-care nurseries throughout the United States. Protracted stays in the nursery are not only costly in economic terms but, in addition, exact a less measurable price from patients, families, and those who provide care. However, the central issue was prognostic. Like the case of Karen Ann Quinlan, this case was clouded by difficulties in concluding that Robin had no "reasonable" hope for survival or that she was neurologically impaired. Although most physicians and nurses agreed to the principle that these were grounds for terminating life support, it was on the issue of her prognosis that they were divided and the social basis of their judgments was laid bare.

The final conference about Robin included a presentation of Robin's history, in which the attending physician and the resident established the parameters of a perplexing prognosis: the late onset of her symptoms, her failure to improve, and her surprising recovery when weaned from the respirator. The evidence cited by both physicians was obtained from diagnostic technology and the physical examination: Nothing was said about Robin's ability to interact. The attending physician was concerned about Robin's neurological status. But after examining evidence from the normal brain scans and "unclear" epidemiological studies of malnutrition, he could not conclude with any "certainty" that Robin was neurologically impaired. Although he suspected that Robin's illness might be irreversible, her remarkable recovery in the past, and the miraculous recovery of another patient with chronic lung disease, gave him reason to doubt his own judgment. Unable to find hard evidence of neurological damage and unable to establish with certainty that Robin's lung disease was irreversible, the attending physician was at a loss as to how to proceed.

The nurses' comments, however, appeared to inhabit an altogether different realm. They expressed "desperation" at having to care for an infant who "cries and screams" and was described as "unsocialized," "unrewarding," possibly "autistic," or "damaged." The attending physician, however, did not consider the desperation of the nurses, and he challenged the scientific status of their comments as representing prognostic signs. When the nurses expressed their frustrations with Robin's stiffness, the attending physician portrayed this condition as a usual response to a medical affliction. When a nurse noted that Robin "likes" to eat, an attribution of emotion, the attending physician countered this with medical information. When a third nurse suggested that Robin might be psychologically damaged or autistic, the attending physician debated this with information from the follow-up study.

The attending physician and the nurses, then, not only articulated contrasting conceptions of Robin as a patient but had differing visions of the central dilemma. For the neonatologist, the dilemma was one of profound prognostic perplexity; for the nurses, the dilemma was one of continuing to care for an infant whose "unsocialized" behavior provided them with few rewards and who, in addition, may have been damaged by their own neglect. The outcome of this conference was a decision to postpone the final decision for a ten-day period. During this time Robin died unexpectedly—making her own decision and sparing the staff from having to make another.

Interviews confirmed what that conference had suggested: that physicians and nurses had radically different views of Robin's prognosis and, thus, different perspectives on the decisions that were made. All of the fellows and most of the residents agreed with the major decisions that were made; most of the nurses disagreed with them, stating that life support should have been discontinued much sooner. Citing the unusual, "and therefore unpredictable," features of Robin's case, residents and fellows felt that Robin might have had a small chance for improvement, and, for this reason, a decision to terminate life support would not have been appropriate. By contrast, more than half of the nurses said that they had "long ago" reached the conclusion that Robin would never recover. Moreover, all but two of the nurses and none of the physicians cited interactive cues that led them to two conclusions. Some felt that Robin was neurologically impaired: "In the end I refused to look after the baby—I couldn't stand her; she just didn't act like a baby. [How did she act?] She would be rigid and she had a real terrible cry, and I don't know, she just sounded to me like she didn't have a brain . . . she never cried for a reason."

Other nurses viewed Robin as unsocialized, autistic, or psychologically damaged from months of life in an oxygen tent or from their own neglect:

> I think they should have stopped on her a long, long time before they did . . . and you can tell me, and they come along and say . . . "this little life can amount to something great." That doesn't . . . make the decision right, because sure, you don't know exactly what's going to happen, but you have a pretty good idea she's going to be damaged. . . . She was psychologically damaged by the time she died. No one loved Robin for her eight months of life. She was handled only when something had to be done to her. And it was her stiffness. Every time you would approach, she would become stiff and withdraw. She could feel the frustration in your hands. Finally, I just said, "Well, I'll just gavage [feed by a tube to the stomach] you and leave you there in a corner." So no one talked to her, no one handled her, she never felt loved. After eight months here I was sure she was psychologically damaged for life.

Taken together, the conference and the interviews suggest that different views of the "facts," "data," and "evidence" can lead to conflicting views of how life-and-death decisions should be made. How may we account for what appear to be different conclusions about Robin's prospects for recovery? The answer, I believe, is organizational. The patient who fails to improve has a different social meaning for the nurse who sits at the bedside for days, weeks, and months than for the physician who cares for the patient during a month-long rotation. For the physician, the survival of another patient provides sufficient grounds for hope; for the nurse, a protracted clinical course provides sufficient grounds for desperation.

When a patient's course is chronic and resists typification and when, in addition, conventional diagnostic technology fails to provide unequivocal evidence of brain damage, conflicting prognostic interpretations are likely to proliferate. At this point, the social origins of these interpretations are revealed. What is at issue is two radically different perceptions of the same patient. One view bases its conclusions on diagnostic technology, physical findings, and epidemiologic studies; in the other view, on social interaction reflecting the perspective of continuous contact. For the nurses, Robin is not merely a patient who violates expectations about a normal course for premature infants with lung disease but is one who violates basic assumptions about normal or appropriate social interaction. The two perspectives meet in the ethics conference, and, in the confrontation between the two modes of knowing, it is no surprise that the former prevails.

Organizations allocate different types of information to the participants in life-and-death decisions. They also allocate different resources to the participants that they use to promulgate their respective points of view and, hence, influence the probabilities that the views of some will prevail. The value of the various types of knowledge closely parallels the stratification of the occupational groups. Despite the fact that nurses can observe certain aspects of a patient's behavior that is not accessible to the other participants, the interactive cues noted by the nurses are *devalued data*. I am not suggesting that these interactive cues have "predictive validity," or that Robin was, "in fact," autistic. Instead, I am suggesting that the knowledge that can be gleaned only through continuous contact be treated as an open question that is subject to further study rather than peremptorily cast out of the realm of medical discourse. A culture that allows only certain types of knowledge to be used as the criterion of certainty may impel some physicians to continue supporting an infant's life long after this may be appropriate.

In the life-and-death decision, facts and values, ideas and interests, and cognitions and affect become inextricably intertwined. The Simpson case illustrates a complex interplay between the perspectives of engagement and detachment. Just as the physicians' structural and emotional disengagement from the patient and the consequences of their decisions may have led them to insist upon absolute certainty, so too, the nurses' pessimism may have been colored by their frustrations in caring for an infant who failed to improve and posed behavioral problems. Continuous contact has its shadow side, permitting emotions, including negative ones, to develop that may compromise the quality of an infant's care.

There seems to be something "wrong" with a set of decisions that adversely affected virtually everyone involved. However, when viewed from within their own frames of reference, the actions of all the participants seem understandable. It is, perhaps, the larger organizational framework that should be questioned.

Conclusions

In this paper I have approached life-and-death decision from the standpoint of the sociology of knowledge. Prognostic conflict in life-and-death decisions was used as a paradigmatic case to illuminate how the organization of the intensive care as an ecology of knowledge may result in conflicts concerning life-and-death decisions. A major finding of my study was that physicians and nurses, because of their differing work experiences, may come to develop conflicting conceptions of the future of infants whose lives are in question. For example, residents base their prognostic assessments largely on data acquired by means of sophisticated measurement instruments (technological cues). My point is not that physicians as individuals assume an uncaring attitude toward their patients, but rather that, structurally and organizationally, they are disengaged from them.

The contrast between physicians and nurses provides a natural experiment, as the social situation of the nurse is, to a certain extent, unique. Nurses are sociologically significant insofar as they, unlike physicians, sustain continuous contact with their infant patients and derive much of their work satisfaction from interaction with those infants who are medically and socially responsive. For this reason, nurses have unique access to a type of information that emanates from such interaction. I have also suggested that this type of knowledge (interactive cues) merits more detailed study. This is not to say, however, that the decisions of nurses are necessarily superior to those of physicians. The Simpson case suggests that the continuous contact between nurses and patients may not be without its shortcomings; that the very structural engagement of nurses and patients may lead nurses to assume a more pessimistic attitude toward infants who are unresponsive, pose behavioral problems, or require chronic care—infants who pose management problems.

Each set of participants, then, approaches the life-and-death decision from a very partial knowledge base. Moreover, the types of prognostic knowledge used in life-and-death decisions are not valued equally. As the data indicate, technological cues and interactive cues do not carry equal weight within the realm of medical discourse; in the confrontation between these modes of knowing, the technological prevails.

These cultural and structural features of technology-intensive medical settings complicate the process of reaching life-and-death decisions. Newborn intensive care is grounded in a culture that accords privileged status to certain types of knowledge and a social structure in which physicians are organizationally disengaged from patients and parents and the task of interaction with the patient is vested largely in those subordinate in the occupational hierarchy. We are left, then, with partial and selective visions of reality, in the sense that Mannheim (1936) intended the term "ideology," and with differential resources with which participants are able to promulgate their respective points of view. Physicians have medical knowledge but may lack potentially valuable sources of information that can be acquired only through interaction with infants. Nurses have access to this more subjective source of information but may be unduly influenced by the practicalities of patient management. As parents become increasingly reliant upon physicians to interpret an increasingly esoteric knowledge, they run the risk of becoming peripheral to life-and-death decisions, and a truly informed consent becomes difficult to attain. Most significant, those who have the most patient contact (the nurses) and the most at stake (the parents) have the least authority in such decisions.

I would also suggest something about how the quality of life-and-death decisions may be improved. To the extent that decisions cannot be extricated from the social organization of the intensive-care nursery, broader changes in that organization may be necessary. A culture that permits the exploration of other modes of knowledge and a social structure that facilitates greater interaction among physicians, patients, and parents may contribute to more informed and equitable decision making.

Acknowledgments

This chapter was prepared with the partial support of the National Institute of Mental Health (USPHS-MH14582; O. Grusky, program director) and the National Institute of Handicapped Research (NIHR-G008006082: O. Grusky, principal investigator), although neither agency is responsible for the views presented. I would like to thank Aaron Cicourel, Fred Davis, Joseph Gusfield, Kristin Luker, Albert Jonsen, and Oscar Grusky for their comments on an earlier draft.

Notes

1. In both nurseries that were studied, parents were somewhat peripheral to the decision-making process—that is, they were consulted in order to elicit their consent to a decision already made by professionals.

2. The data were collected during sixteen months of field research in two intensive-care nurseries that contrasted in size, prestige, referral patterns, and the demographic composition of their clientele. One setting, which I call Randolph, is a twenty-two bed nursery in an elite institution; it functions as a referral center and serves a demographically heterogeneous clientele. The other, a sixty-bed nursery, does not serve as a referral center and serves a largely indigent clientele. The major methods of data collection were observations of life-and-death decisions and fifty-eight semistructured interviews with physicians; with a random sample of nurses, stratified by age, experience, and shift; and with parents.

3. This discussion pertains to normative ethics rather than to metaethics. Normative ethics is concerned with the values and principles that characterize right or wrong conduct. Metaethics is that branch of philosophy that examines the nature of moral discourse. Normative ethics has been far more significant to the formulation of policies concerning decisions in newborn intensive care.

References

Anspach, Renee R. 1987. Prognostic Conflict in Life-and-Death Decisions: The Organization as an Ecology of Knowledge. Journal of Health and Social Behavior 28, no. 3 (September):215–231.

Blalock, Hubert M. 1960. Social Statistics. 1st edition. New York: McGraw-Hill.

Budetti, Peter, Peggy McManus, Nancy Barrano, and Lu Ann Heinen. 1980. The Cost-Effectiveness of Neonatal Intensive Care. Washington, D.C.: United States Office of Technology Assessment.

Crane, Diana. 1975. The Sanctity of Social Life. New York: Russell-Sage.

Foucault, Michel. 1975. The Birth of the Clinic: An Archaeology of Medical Perception. New York: Vintage.

Fox, Renee. 1974. Ethical and Existential Developments in Contemporary Medicine. Milbank Memorial Fund Quarterly 54:445–483.

Fox, Renee, and Judith Swazey. 1973. The Courage to Fail. New York: Dodd-Mead.

Guilleman, Jeanne Harley, and Lynda Lytle Holmstrom. 1986. Mixed Blessings: Intensive Care for Newborns. New York: Oxford University Press.

Janis, Irving, and Leon Mann. 1977. Decision Making. New York: Free Press.

Jonsen, Albert, and George Lister. 1978. Newborn Intensive Care: The Ethical Problems. Hastings Center Report 8:15–18.

Mannheim, Karl. 1976. Ideology and Utopia: An Introduction to the Sociology of Knowledge. New York: Harcourt-Brace.

Parsons, Talcott. 1951. The Social System. New York: Free Press.

Reiser, Stanley Joel. 1978. Medicine and the Reign of Technology. Cambridge: Cambridge University Press.

Wilson, Thomas. 1970. Normative and Interpretive Paradigms in Sociology. Understanding Everyday Life, Jack Douglas, ed., pp. 111–172. Chicago: Aldine.

The author acknowledges on p. 54, line 11 the work of Guilleman and Holmstrom (1986).

5

Deciding to Treat Newborns: Changes in Pediatricians' Responses to Treatment Choices

Jeanne Harley Guillemin, I. David Todres, Dick Batten, and Michael A. Grodin

Only since the late 1970s has medical treatment for newborn patients ceased to be a matter of special case analysis and become the focus of widespread public controversy, including federal regulations. The growing complexity of both the technology and the organization of hospital care for newborns accounts for most of this change. A few elite intensive-care nurseries instituted in teaching hospitals during the 1950s have turned out to be the models for services in virtually every large hospital in the United States, as well as for many foreign hospitals. Case numbers are now significant; at least two hundred thousand newborns in the United States each year are referred to intensive care. Consequently, many decisions in the category "lifesaving" call for a rational policy rather than the private interpretations of physicians and parents (Guillemin and Holmstrom 1986).

With the proliferation of medical options for the newborn patient, physicians—in particular, pediatricians—face new dilemmas in selecting appropriate treatment. As in adult medicine, difficult clinical choices sometimes must be made as to whether an infant patient survives with impairment or has treatment withheld to avoid a life of suffering or irreversible coma. Yet, in contrast to adult cases, there is no estimate that can be made of the infant's willingness to hazard medical risks or to choose between prolonged suffering or death. Instead, authority has been traditionally vested in parents to make such decisions in consultation with a physician (President's Commission 1983).

Early ethical debate about infant medical care focused primarily on the newborn with severe congenital anomalies whose life could be saved but whose condition could not be corrected by medical intervention. The Down's syndrome infant with surgically remedial intestinal blockage is one case in point (Gustafson 1973); the infant with spina bifida is another (Swinyard 1978). Attention has now turned to the low-birthweight (weighing less than 2,500 grams, or 5.5 pounds) infant suffering from prematurity, the major cause of newborn mortality in the United States (McCormick 1985; Institute of Medicine 1985). These infants, more than those with single anomalies, represent the majority of infants in newborn intensive care. The treatment dilemmas posed by critically ill newborns have prompted a variety of responses from hospital personnel and administrators (Levin 1985; Grodin, Schwartz, and Todres 1984). Many hospitals, for example, sponsor some kind of ethics committee—ad hoc or formal—as a forum for discussing problem cases.

Intervention by agencies outside the hospital has also increased in recent years. In several well-publicized instances, the courts have intervened to pass judgment on the medical treatment of newborns with congenital anomalies, including a Down's syndrome infant with esophageal atresia (narrowing of the esophagus), an infant with spina bifida, and Siamese twins. The most noteworthy of these court cases, the Indiana Baby Doe case, precipitated federal action aimed specifically at medical decision making for newborns. Drawing on existing law that protects the rights of the handicapped, the U.S. Department of Health and Human Services in 1982 issued the Baby Doe regulations to all hospitals receiving federal funding (Annas 1983). The federal government's initial approach was direct intervention, via publicized telephone "hot lines," and official demands for medical records in specific case decisions. Subsequent court and congressional action has shifted emphasis away from the handicapped law to laws pertaining to child abuse and neglect and to the role of existing state agencies in protecting infant patients (Rhoden and Annas 1985). At the hospital level, the federal government urges, though does not legally require, the creation of infant-care review committees to play an active role in arbitrating clinical decisions and, if necessary, in alerting state agencies to possible infant abuse.

The current Baby Doe regulations remain strongly protective of the right of newborn patients to medical care, ruling out possible long-term disability as a valid reason to withdraw care. A broad coalition of medical professionals, hospital organizations, social welfare groups, and antiabortion groups endorsed the protective purpose of the regulations, stressing that "in cases where it is uncertain whether medical treatment will be beneficial, a person's disability must not be the basis for a decision to withhold treatment" (American Academy of Pediatrics 1983:560). This belief is echoed in the regulations, which essentially disallow cessation of treatment except when the infant patient is dying or irreversibly comatose.

Despite changes in the organization and technology of hospital care for newborns, physicians retain the discretionary authority to assess medical need and treatment benefits for the infant patient. Even in tertiary care centers, where teamwork is essential, pediatricians set the clinical standards and determine the course of therapy. At the local level, pediatricians are often involved in referral decisions concerning borderline cases. The attitudes of practicing pediatricians are, therefore, crucial to the implementation of regulations and will shape the consequences of efforts to protect the newborn patient.

Research Methods

In a survey conducted in 1975, Massachusetts pediatricians were asked to indicate their judgments concerning two problem clinical cases and, further, to express their values concerning the resolution of ethical dilemmas in infant care (Todres, Krane, and Howell 1977). The sample consisted of 406 physicians designated as pediatricians in the 1972–1973 Massachusetts Datawell Directory. Of these, 230 (57 percent) responded, an adequate-to-good return rate to draw inferences to the overall population (Babbie 1979). Respondents were asked to make treatment versus no or limited treatment recommendations for two newborn cases. The first was a Down's syndrome infant with duodenal atresia (narrowing of the duodenum). The second, also a hypothetical case, was an infant with spina bifida.

In 1984–1985, we initiated a follow-up survey, substituting the mailing list of the Massachusetts branch of the American Academy of Pediatrics for the out-of-print Datawell source. Recognizing medical and regulatory developments over the last decade, we expanded the original questionnaire in several ways. One was to include a third clinical case involving therapy for a very-low-birthweight newborn with asphyxia (lack of oxygen in the lungs). Questions relating specifically to the Baby Doe regulation were also added, along with inquiries about the potential function and composition of hospital bioethics committees. Variables such as religious activity and religious affiliation, measured at the nominal level, have been statistically dichotomized to be treated as interval variables. Of a sample population of 801 (discounting those retired, deceased, and relocated), 449 (56 percent) of the pediatricians queried in 1984–1985 responded. The significance of responses was calculated using the Pearson's correlation coefficient. Only the correlation coefficients were reported that were found to be significant at the .05 level.

A number of the population characteristics of the 1984–1985 sample are notable. Compared to 1975, the proportion of women pediatricians responding increased, reflecting the generally greater representation of women in pediatrics. The percentage of women respondents moved from 12 percent to 27 percent. The median age for respondents dropped from 44 in 1975 to 41 years in 1984–1985. The median age for women respondents in 1984–

1985 was thirty-years years; for men, forty-four years. The majority religious affiliation of male respondents in 1975 and 1984–1985 also differed. A decade ago Catholic practitioners were 27 percent of the sample, Jewish practitioners 25 percent, and Protestant practitioners 24 percent. In the current survey, Jewish physicians constituted 37 percent of the respondents, Protestants 26 percent, and Catholics 24 percent. Other characteristics of the 1984–1985 sample of physicians returning questionnaires included a high percentage of married pediatricians (87 percent) and of board certification in pediatrics (87 percent). Many responding physicians were graduated from prestigious medical schools, 67 percent from institutions ranked in the top two categories of the 1980 Gorman Report.

Nearly 60 percent of the respondents were actively attending critically ill infants. Forty percent of the pediatricians had treated between five and twenty such patients in the previous twelve months. Another 19 percent treated more than twenty critically ill newborns per year.

Research Findings

Case I

In the first case, physicians were asked to indicate whether they felt an infant with Down's syndrome should be operated upon for duodenal atresia, despite objections by the parents. In 1975, 46 percent of the respondents would have chosen to operate despite parental objections. In 1985, 73 percent of the pediatricians would have operated. In 1975, 40 percent of the respondents who recommended surgery indicated that they would seek a court order if parents persisted in their objection. In 1985, 68 percent of the pediatricians indicated they would seek a court order.

As in 1975, religious affiliation was associated with the decision to operate. Catholic pediatricians were more likely than Protestant or Jewish physicians to choose the operation (see Table 5.1). However, compared to 1975, proportionally more Protestant and Jewish physicians would operate, with apparently the greatest change in attitude among Jewish practitioners. Women pediatricians were also more in favor than men of the duodenal atresia operation. This is in contrast to the 1975 result, which found women less in favor of the operation than men.

In the current survey, younger physicians were more likely to choose the operation than were older practitioners, as were physicians from lower-ranked and foreign medical schools. Younger physicians were also more likely to resort to a court order. Pediatricians who were married, those who attended relatively large numbers of critically ill newborns, and those who came from lower-ranked or foreign medical schools were also more likely to seek a court order over parental objection.

TABLE 5.1

1975 and 1985 Responses, Case I, by Gender and Religion

	Percent of Total		Percent Recommending Surgery	
	1975	1985	1975	1985
Female	11	26	29	79
Male	89	74	50	72
Catholic	27	27	66	85
Protestant	24	25	50	69
Jewish	25	35	38	69

Asked what else they would do to deal with parental objections, 68 percent of respondents said they would personally try to persuade the parents to consent to the surgery. Another 25 percent said they would bring in others—social workers, psychologists, or the clergy—to convince parents.

Case II

In the second clinical case, physicians were asked to consider three options in treating an infant born with severe spina bifida, weighing 3.15 kilograms (6.5 pounds) and with a small head. The first option was "no surgery but . . . [with] skilled nursing treatment including an incubator, tube feeding, and antibiotic[s]. . . ." The second option was "no surgery, . . . [with only] custodial care, ordinary feeding and nursing procedures, but no more." The third option was an emergency operation on the spinal lesion.

In contrast to 1975, when only 33 percent of respondents recommended emergency surgery, 53 percent did so in 1985. In 1975, physicians under fifty years of age recommended surgery more often than those age fifty and older. In 1985, younger physicians were also more willing to recommend surgery. In 1984, married physicians recommended surgery more often than did unmarried physicians.

Of the pediatricians advocating surgery in this case in 1985, 56 percent indicated that they would change their choice if parents withheld consent. In 1975, 60 percent of the pediatricians opting for surgery would have changed their minds if parents refused surgery. On the other hand, physicians not initially in favor of surgery indicated willingness to recommend surgery if parents asked them to "do everything possible." In 1975, 74 percent of those opposing surgery would have changed their decisions; in 1985, 64 percent of those opposed would change their decisions to accommodate parents.

Older and Catholic physicians indicated they would be less likely to change their decisions to operate if parents withheld consent.

Case III

In the third, and new, clinical case, physicians were asked whether they would treat a 700-gram (1.5-pound) infant, born at twenty-eight weeks gestation and suffering from asphyxia at birth.

The great majority (90 percent) of respondents indicated that resuscitation should be continued and the infant transferred to a neonatal intensive-care unit. Once again, age influenced the treatment decision; younger physicians were more in favor of resuscitation and transport than were older colleagues. Physicians treating many critically ill infants also tended to be more in favor of continued therapy.

For 65 percent of the respondents, parental objections to this treatment would not affect their decision. Catholic physicians, married physicians, and those working with high case volumes of newborn patients were most likely to hold to the decision to treat. Women pediatricians, however, were more responsive than their male colleagues to parental requests not to treat in this case.

Queried further about the viability of very-low-birthweight infants, many respondents (71 percent) indicated that even a 600-gram (1.3-pound) newborn with asphyxia should be revived and transported. Respondents also indicated, however, a diversity of reasons for this decision. Twenty-four percent felt that this infant's chances for survival were good. Another 20 percent were unsure about viability and favored transport as a means of buying time. Fifteen percent felt that the infant's survival chances were bad; of these, some (6 percent) recommended transport anyway.

Queried about other types of low-birthweight cases, most physicians (97 percent) felt that the 600-gram newborn not suffering from asphyxia should be given intensive care. Likewise, many physicians (87 percent) felt that the 700-gram infant was not an experimental case. One woman pediatrician, graduated from medical school in 1980, wrote this comment: "When I began my training, 700-gram infants were considered 'non-viable.' Now enough are surviving without major sequelae to make at least the initial effort worthwhile. How many would now survive if we didn't at least try over the last few years?"

Physician Values

The clinical decisions that physicians make often have broad repercussions for the parents who might bear the burden of grief or long-term care of a handicapped child and also for entire systems of health and social services. The second part of the questionnaire addressed these issues.

Asked, "Should the family's financial situation enter into your decision to forego life-sustaining treatment?" a great number (93 percent) of physicians indicated it would not. Even more (95 pecent) felt the family's lack of medical insurance would not influence their decisions. A majority (64 percent) felt that the lack of rehabilitation services for handicapped infants would not affect their treatment choices in difficult cases.

As in 1975, pediatricians were asked if "the presence of mental retardation or severe physical malformation [would] justify withholding consent for lifesaving procedures for psychosocial reasons"—for example, a negative effect on the parents' marriage or a burden on the family. A decade ago, Massachusetts pediatricians were evenly divided on this matter, with half indicating such factors would influence their decisions and half indicating they would not. In 1984, most pediatricians (64 percent) indicated that possible psychosocial consequences are not a reason to withhold treatment. Catholic pediatricians and religiously active practitioners were especially likely to take this position. Physicians treating high volumes of critically ill infant patients also indicated that psychosocial repercussions would not justify withholding treatment.

Concerning future developments in medicine, Massachusetts pediatricians in 1984–1985 gave a strong vote to research on the newborn intensive-care-unit patient (98 percent). Asked if such research would be acceptable if there were no immediate benefits to that child but future infants might benefit, 94 percent of the responding physicians indicated their approval. At the same time, most (61 percent) indicated that, with limited economic resources, top priority should be given to the development of better prenatal care.

In 1975, 32 percent of respondents indicated that, under certain prescribed circumstances, an infant euthanasia law should be allowed. In 1985, 28 percent indicated this opinion, especially physicians under fifty years of age.

Attitudes Toward Baby Doe Regulations

Virtually all responding physicians (98 percent) indicated familiarity with the Baby Doe regulations. Many (86 percent) felt that the regulations were too intrusive. Asked what treatment constraints the regulations impose, 82 percent responded that to the best of their knowledge, the regulations demanded surgery for a Down's syndrome infant with duodenal atresia.

Many (66 percent) believed that the regulations require maximum therapy for the severely premature infant; 27 percent saw the regulations as requiring maximum therapy for an infant with anencephaly. Most (62 percent) of the responding pediatricians rejected the physician and parents as sole decision-makers in problem cases. A great number (86 percent) believed that the nurse member of the clinical team should have input in decision making. On an institutional level, 70 percent of the pediatricians reported that the hospitals with which they were affiliated had no specific policies for treatment of critically ill and defective newborns.

Since 1975, physician support for hospital ethics committees appears to have increased. In the first survey, only 32 percent felt that a committee of any sort could effectively function to support the responsible physician's decision; most of the respondents favored no committee involvement at all. In the 1984–1985 survey, 62 percent of the respondents indicated their approval of a hospital review committee. However, 84 percent of the pediatricians saw the role of the review committee as advisory rather than prognostic or related to quality assurance. Asked who should be on a hospital review committee, respondents were generally in favor of the membership suggested by the federal government for infant-care review committees but had varying levels of enthusiasm. The strongest endorsements went to the consulting specialist (99 percent), the primary physician (98 percent), and the nurse (94 percent), with the least support (42 percent) for representatives of disability groups.

Increase in Favor of Medical Intervention

A major finding of this study is the new and more positive pediatrician evaluation of the viability of newborn patients. The data suggest multiple reasons for this change. For example, the treatment of very-low-birthweight newborns is generally supported as frontier area in pediatric medicine. As most newborn intensive-care patients are in this category, strong support for research on the newborn intensive-care-unit patient suggests professional consensus that this is the cutting edge of the profession. Although follow-up studies on infants born at less than 1,000 (2.2 pounds) grams have indicated high rates of serious morbidity (see Budetti, McManus, and Barrano 1981; Kitchen and Murton 1985; Brinton, Fitzhardinge, and Ashley 1981; McCormick 1985), most practitioners seem optimistic about improved treatment outcome. Belief in medical progress based on survival data alone, therefore, increases the positive estimate of the viability of infants born prematurely or small for gestation age. However, a change in the value put on infant life, rather than faith in medical progress, might better explain the more positive professional attitude toward the Down's syndrome infant, as the corrective procedure for the intestinal blockage associated with this condition has been and remains simple.

As in 1974–1975, religious affiliation appears to influence treatment decisions. In the Down's syndrome case in particular, Catholic physicians opt more for surgery and reaffirm the value that mentally handicapped or deformed infants should be saved. According to Crane's discussion of religious influence on pediatric decisions (Crane 1977), Catholic physicians might be expected to place a special value on the life of the newborn, more than their non-Catholic colleagues. Yet, it is also true that the attitudes of all physicians have come into greater alignment concerning lifesaving treatment for Down's syndrome infants. At present, the distinction between Catholic and other practitioners on this issue is considerably less than a decade ago. Greater awareness of the potential of such handicapped infants and of the increased services for their education have perhaps influenced this change. A more protreatment attitude may also be part of the current medical culture, which advocates emergency lifesaving intervention for adults and children alike and tends to relegate subsequent social problems to other professionals and to the state.

The relatively young age of the current sample of physicians might also have had an influence on increased optimism about infant patient viability. Familiarity with the latest medical techniques might increase the likelihood of intervention by younger physicians, and, conversely, a lack of this familiarity could deter older physicians from treating. In 1975, 61 percent of the respondents were under fifty years old; in 1984–1985, 72 percent of the respondents were less than fifty years old. Today, younger physicians appear almost uniformly protreatment, whether the issue is corrective surgery or transport to intensive care. In 1975, younger physicians were also more likely to recommend surgery than were their colleagues over fifty.

The trend toward a more protreatment attitude is also indicated in the difference in women pediatricians' attitudes in 1975 compared with 1984–1985. Ten years ago, women physicians were less likely than men to recommend surgery in the Down's syndrome case and just as likely as their male colleagues to opt for surgery for the spina bifida infant. Now women in practice seem more likely to recommend surgery in the first case and just as likely as men to opt for treatment in the second and third cases in the study. An exception to this conformity appears to be the greater responsiveness of women physicians to parents who do not wish treatment for the 600-gram infant with asphyxia. Still, the reasons for this difference are unclear and the question requires further study.

Diminished Parental Influence

Although several physicians in the study added comments about their concern with the social consequences of treatment decisions for families, the influence of parents on these decisions appears to have diminished. Increased physician willingness to use a court order in the case of a Down's syndrome infant with duodenal atresia is one indication of this change. Slightly increased resistance to parental wishes, as shown in the responses given for the second case, is another example. Increased rejection of psychosocial reasons for withholding lifesaving treatment is still another.

One way to explain this shift in attitudes is that physicians have become more aware of the rights of the newborn as they might conflict with parental rights and also more protective of the newborn's interests in isolation from the family. An alternative explanation is that the expansion of intensive-care services for newborns automatically puts treatment decisions in a bureaucratic context and consequently tends to depersonalize relations between parents and practitioners. Several studies document this tendency (Bogdan, Brown, and Foster 1982; Sosnowitz 1984). The routine nature of neonatal referral raises the issue not only of diminished parental control but of appropriate institutional safeguards of the infant patient's interests.

Interpretation of Regulations

The stringency of the current Baby Doe rule has been commented on by Rhoden and Arras (1985) to the effect that it mandates complex treatment even in cases such as infants with Tay-Sachs disease or Trisomy-13 (in which an extra, thirteenth, chromosome is present). Physician attitudes, as represented in this study, are generally protreatment and therefore consonant with present government policy. But there are several divergences. For one, although most practitioners see the regulations as requiring routine surgery for the Down's syndrome newborn, not all physicians express a willingness to choose treatment in this case. At the other extreme, more than a quarter of the respondents incorrectly understood the regulations to require treatment of a hopeless case—that is, the anencephalic newborn.

By coincidence, the second clinical case example, that of the infant with spina bifida, replicates the 1983 Baby Jane Doe legal case, which tested the limits of federal authority in applying handicapped protection law to infants with severe congenital anomalies. The Superior Court of New York and the Supreme Court have upheld the lower court ruling that the law did not require this infant to have surgery. Still, many physicians recommended surgery for this hypothetical case.

Although pediatricians once appeared firmly committed to quality-of-life criteria in assessing borderline cases, the responses in our study suggest a greater emphasis on the newborn's options for medical therapy and on the immediate positive results of treatment. In their 1977 report on a national study of pediatricians, Shaw, Randolph, and Manard found that nearly 90 percent of the respondents ranked "potential quality of life" as a major factor in making nontreatment decisions. Our study suggests that physicians now rely less on this criterion.

The Role of the Committee

The mediation of a disinterested third party—neither the parents nor the physician—has frequently been advocated to resolve cases in which the infant patient's rights are potentially not represented (Weir 1984). The hospital ethics committee, in particular, has in recent years evolved as an important potential solution to this problem. In its survey, the American Academy of Pediatrics (1984) found that of 426 hospitals having special-care pediatric units, 57 percent had infant-care review or bioethics committees. Of the hospitals without committees, 75 percent (132) are considering instituting such groups.

Both the academy survey and one done for the President's Commission (Younger 1983) indicate that the activities of existing hospital committees are primarily educational forums for discussing cases after the fact. In contrast, the model committee advocated by the federal government would have more central involvement in clinical decisions, including the power to override either physician or parental authority should the committee's opinion be different from that of either or both of these decision-makers.

The role ascribed to the hospital ethics committee by respondents to this questionnaire matches existing committee activities rather than authoritative functions advised by the government for infant-care review committees. Respondents supported the advisory function of existing hospital committees as well as their drawing membership from the professional community. Physicians, nurses, hospital administrators, social workers, and lawyers are frequently listed as members of existing hospital review committees and were also generally approved by the 1984–1985 physician respondents.

Advocates for the disabled are suggested as infant-care review committee members by the federal government but are not in evidence on existing boards. This might be the result of practical circumstances; that is, it may be more convenient to call a quick meeting of hospital personnel and involved parents than to recruit extra-hospital participation. Or resistance to representatives for the disabled may also indicate resistance to any external extra-hospital influences on decision making. In either case, it remains to be seen if hospital committees will be effective means for protecting the infant patient's interests or simply a vehicle for enhancing professional input.

The increased favor with which this group of physicians regards committees could indicate that the reported lack of policy concerning newborn patients will eventually be addressed. Hospital committees have the potential for educating personnel about ethical and legal issues and formulating policy guidelines. It is clear from our research that physicians are more willing cases. How parents could or should have input into treatment policy remains an unanswered question.

Appendix

Case I

Baby A is a Down's infant born with duodenal atresia. The parents have been consulted and do not want to consent to surgery to repair the intestinal obstruction. Do you feel this baby should be operated on?

Yes ____ No ____

If yes, would you:

A) Obtain a court to operate over the parents' refusal?
Yes ____ No ____

B) What else would you do to solve this dilemma?
____

If no, would you:

A) ____ Give custodial care and ordinary nursing procedures?

B) ____ Speed up the death process by administering barbiturate or any other drug in toxic doses.

C) ____ Other:
____

Case II

Baby B was born with a large thoraco-lumbar meningomyelocele. The baby's weight is 3.15 kg., head circumference is 32 cm. (3rd percentile) and the baby is paraplegic and incontinent. It is anticipated that there is a very high risk of hydrocephalus developing. What treatment would you recommend?

____No surgery but would provide skilled nursing treatment including an incubator, tube feeding, and antibiotic therapy.

____No surgery, would provide custodial care, ordinary feeding and nursing procedures, but no more.

____Would recommend emergency operation on back lesion.
If the parents withhold consent for surgery, would this change your choice for emergency operation? Yes ____ No ____
If the parents ask you to do everything possible to preserve the life of the child, would this change your choice for surgery?
Yes ____ No ____

Case III

Baby C is born precipitously at 28 weeks gestation and weighs 700 grams. Because of asphyxia suffered by the infant at the time of delivery, the infant has required vigorous resuscitation in the delivery room. What further treatment would you recommend?

____Continue intensive resuscitative measures and transfer the infant to a neonatal intensive-care unit.

____Discontinue resuscitative measures and give comfort care (warmth and gavage feedings).

If the parents ask you not to continue these resuscitative measures on their infant, would this change your above choice for treatment?

Yes ____ No ____

If the infant's birthweight was 600 grams with a similar clinical history, would this change your recommended treatment?

Yes ____ No ____

If the infant's birthweight was 600 grams but did not suffer an asphyxial insult, would you treat and transport?

Yes ____ No ____

References

American Academy of Pediatrics. 1983. Joint Policy Statement: Principles of Treatment of Disabled Infants. Pediatrics 73:559–560.

______. 1984. Survey on Infant Care Review Committees. American Academy of Pediatrics Meeting, September 12 (mimeograph).

Annas, George J. 1983. Disconnecting the Baby Doe Hotline. Hastings Center Report 13:14–16.

Babbie, E. R. 1979. Survey Research Methods. 2d edition. Belmont, California: Wadsworth Publishing Co.

Bogdan, Robert, Mary Alice Brown, and Susan B. Foster. 1982. Be Honest but Not Cruel: Staff/Parent Communication on a Neonatal Unit. Human Organization 41:6–16.

Brinton, S., P. M. Fitzhardinge, and S. Ashley. 1981. Is Intensive Care Justified for Infants Weighing Less than 801 Grams at Birth? Journal of Pediatrics 99:937–943.

Budetti, Peter, Peggy McManus, and Nancy Barrano. 1981. Case Study 10: The Costs and Effectiveness of Neonatal Intensive Care. Washington, D.C.: Office of Technology Assessment.

Crane, Diana. 1977. The Sanctity of Social Life. New Brunswick, New Jersey: Transaction Books.

Datawell's 1982–83 Medical Directory of Massachusetts. 1983. West Roxbury, Massachusetts: Datawell Associates.

The Gorman Report. 1983. Los Angeles: National Educational Standards.

Grodin, Michael, W. Markely, and A. McDonald. 1984. A Team Approach to Medical-Legal and Ethical Dilemmas in a Teaching Hospital. Institutional Ethics Committees and Health Care Decision-Making, R. Cranford and A. Duodera, eds., pp. 118–128. Ann Arbor: Health Administration Press.

Grodin, M., S. Schwartz, and I. D. Todres. 1981. Moral Dilemmas and Problematic Decision-Making in Newborn Intensive Care. The Rights of Children, J. Henning, ed. Springfield, Illinois: Charles C. Thomas.

Guillemin, Jeanne H., and Linda L. Holmstrom. 1986. Mixed Blessings: Intensive Care for Newborns. New York: Oxford University Press.

Gustafson, J. 1973. Mongolism, Parental Desires, and the Right to Life. Perspectives in Biology and Medicine 16:548–552.

Institute of Medicine. 1985. Preventing Low Birthweight. Washington, D.C.: National Academy Press.

Kitchen, William H., and Laurence J. Murton. 1985. Survival Rates of Infants with Birth Weights Between 501 and 1,000 grams. American Journal of Diseases of Children 139:470–471.

Levin, Betty W. 1985. Consensus and Controversy in the Treatment of Catastrophically Ill Newborns. Which Babies Shall Live? T. Murray and A. L. Kaplan, eds. Clifton, New Jersey: Humana Press.

McCormick, Marie. 1985. The Contribution of Low Birth Weight to Infant Mortality and Childhood Morbidity. New England Journal of Medicine 312:82–90.

President's Commission for the Study of Ethical Problems in Medicine and Biomedical and Behavioral Research Deciding to Forego Life-Sustaining Treatment. 1983. Washington, D.C.: U.S. Government Printing Office, pp. 197–229.

Rhoden, N., and J. Arras. 1985. Withholding Treatment from Baby Doe. From Discrimination to Child Abuse. Millbank Memorial Fund Quarterly/Health and Society 63:18–51.

Shaw, A., J. Randolph, and B. Manard. 1977. Ethical Issues in Pediatric Surgery: A National Survey of Pediatricians and Pediatric Surgeons. Pediatrics 60:588–599.

Sosnowitz, B. 1984. Managing Parents on Neonatal Intensive Care Units. Social Problems 31:390–402.

Swinyard, C. A. 1978. Decision Making and the Defective Newborn. Springfield, Illinois: Charles C. Thomas.

Todres, I. David, D. Krane, and M. C. Howell. 1977. Pediatricians' Attitudes Affecting Decision-Making in Defective Newborns. Pediatrics 60:197–201.

Weir, Robert. 1984. Selective Nontreatment of Handicapped Newborns. New York: Oxford University Press.

Youngner, S. 1983. A National Survey of Hospital Committees. Deciding to Forego Life-Sustaining Treatment, pp. 443–466. Washington, D.C.: President's Commission.

6

Decision Making About Care of Catastrophically Ill Newborns: The Use of Technological Criteria

Betty Wolder Levin

Neonatal intensive care enables the treatment of many babies born catastrophically ill. The benefit of providing such treatment is usually clear, for most of these babies grow up to enjoy a good quality of life. For some babies, however, decisions are made to withhold treatment when treatment might lead to a prolongation of the dying process or survival with a very poor quality of life. In this chapter, I explore the way in which technology and conceptualizations about technology influence decisions. In particular, I examine categories such as "ordinary" and "extraordinary" as they are used in discussions of treatment choices. I raise two questions: Why are technological criteria considered in making decisions about the care of newborns? And, are technological criteria relevant for making appropriate treatment choices?

Treatment of the Critically Ill in the Western Medical Tradition

In Western biomedicine, clinical behavior has been guided by ethical traditions based on principles stating that physicians should respect the "sanctity of life" and "do no harm." The clinical interpretation of these principles has changed as medical technology and social values have changed.

When medicine had a more limited ability to cure, an important part of the doctor's role was to help patients accept death. Medical ethics dictated that physicians should not prolong the dying process. With the dramatic reduction in mortality from infectious diseases and the development of many new surgical and medical procedures during the first half of the twentieth century, these attitudes changed. In the years following World

War II, dramatic strides were made in the ability to treat the critically ill, including the development of intensive care for adults and newborns. Death was defined as the enemy and a "technological imperative" developed that dictated that all that could be done should be done to prolong life (Aries 1982; Reiser, Dyck, and Curran 1977).

By the early 1960s, however, some "halfway technologies"—such as the use of kidney dialysis and respirators—that could maintain life but not cure the underlying condition were seen to lead to a cultural contradiction. Clinicians felt that some treatments would result in a conflict between the norm to "preserve life" and the admonition to "do no harm" or "not prolong dying."

To deal with the contradiction, two changes took place. First, there was a change in the definition of death. The traditional definition, which depended on the absence of respiration and heartbeat, became problematic when cardiac and respiratory resuscitation were developed. A new physiological criterion, the absence of brain function, was substituted as a means of defining death (President's Commission 1981). Second, changing attitudes of clinicians and other members of society no longer upheld the absolute value of preserving life. Instead, it became acceptable to weigh other values against the value of life in making treatment decisions (Parsons, Fox, and Lidz 1972). This led to a shift away from a physiological definition of life toward one based on social criteria—in particular, the ability to interact meaningfully with others (Crane 1975). There is now consensus among most clinicians and bioethicists that some treatments may not be mandatory in some circumstances. For example, life support may be discontinued when patients are diagnosed as "brain dead." Controversy continues about many other treatment practices, such as when to stop tube feedings for a comatose patient (President's Commission 1983; Capron and Cassell 1984).

Norms for the treatment of newborns followed the same pattern as treatment for older patients. Prior to the middle of this century, little could be done to save impaired newborns; since that time, many dramatic advances in medical technology and practice have enabled medical professionals to aggressively treat severely impaired newborns. At first, technical criteria alone were used to choose treatments that would prolong survival. By the mid-1960s, babies with severe problems were admitted to newly developed intensive-care units. In the late 1960s and early 1970s, after observing high rates of impairments for some types of infants, many care givers began to advocate the use of social as well as technological criteria in determining treatment choice (Reich 1978; Silverman 1980; President's Commission 1983; Weir 1985; Guillemin and Holmstrom 1986).

To reduce both the emotional stress and cognitive complexity of making treatment choices about catastrophically ill infants, norms have developed to guide such decisions. Among most clinicians and nurses who work in neonatal intensive-care units, there is general agreement about many treatment choices, such as withholding heroic treatments from anencephalic infants. However, controversy continues about other treatment choices, such as performing lifesaving surgery on critically ill infants with severe impairments (Guillemin and Holmstrom 1986; Weir 1985).

The Clinical Model

The following model reflects the clinicians' conceptualizations of treatment choice. It was developed from research on neonatal decision making covering a period of nine years, including surveys and participant observation in a neonatal intensive-care unit (Levin 1986). The model illustrates how clinicians decide which treatments to give if the benefit of treatment is questioned.

The clinician's decision-making process starts with the categorization of patient characteristics along a number of dimensions. These dimensions are assigned different labels, but they may be understood by using the following categories: (1) quality of life; (2) uncertainty; (3) the nature of the critical condition; (4) and social value. The quality of life balances positive attributes such as pleasure and satisfaction with negative aspects such as pain and sadness that may be experienced by the baby. Uncertainty pertains to the lack of ability to predict the prognosis for the infant. The nature of the critical condition refers to the medical problem(s) of the infant that, if not treated, will lead to death. Social value refers to the benefits and burdens to others that will result from sustaining the life of the infant or from allowing the infant to die.

Most discussions concerning withholding treatment from catastrophically ill infants are about which patients should or should not be treated, as if the decision involved a choice between *treatment* and *nontreatment* (Reich 1978). However, in a modern hospital setting, a patient is almost never "not treated" in the sense that *no* treatments are given. Rather, decisions are made about *which* treatments to give and which to withhold from a range of treatment possibilities. Therefore, the important questions include not only Who should be treated? and Who should decide? but also Which treatments should be given and which should be withheld?

As clinicians both give and withhold different treatments from the same patient, they obviously make distinctions based not only on characteristics of patients but also on characteristics of treatments. Treatments are categorized according to: (1) aggressiveness; (2) ordinary and extraordinary care; (3) withholding and withdrawing treatment; and (4) active and passive euthanasia. Aggressiveness is a concept used by clinicians in referring to treatments that have a large physiological effect; are complex, experimental, invasive, and expensive in terms of staff time or monetary costs; and/or involve the use of high technology. For example, treatments using a respirator or neurosurgery are considered to be more aggressive than treatments such as tube feedings and giving antibiotics. Withholding treatment refers to not starting a new treatment, whereas withdrawing treatment refers to stopping the use of an ongoing treatment. Active euthanasia causes death, and passive euthanasia allows death to occur. The distinction between ordinary and extraordinary treatment will be discussed in the next section.

Each of these dimensions is sometimes used dichotomously, yet they each may be seen as forming a continuum. Categorizations along each of the dimensions are based on objective, technical criteria, yet they also involve evaluations using subjective, culturally defined criteria as well. Subjective evaluations may be most obvious in categorizations defining a "good quality of life." "Good quality of life" is a value judgment and may mean different things to different people. Culturally defined criteria also affect evaluations of other categorizations, such as the distinction between ordinary or extraordinary treatments.

Although there is agreement among most clinicians that all of these dimensions may be addressed in decision making, clinicians differ in their opinions about how particular patients and treatments are categorized and in how important each of the dimensions should be in particular cases. Choices are made to provide treatments that are seen as commensurate with patient characteristics. For example, when it is felt that a baby could enjoy a good quality of life, aggressive treatments will be provided. However, when clinicians feel that aggressive treatments would lead to a very poor quality of life, they may withhold treatments that they feel are extraordinary. Although there is growing consensus that it is acceptable to withhold extraordinary treatments, clinicians differ about which treatments they consider extraordinary in particular cases.

Categorizing Treatments as Ordinary or Extraordinary

The distinction between ordinary and extraordinary treatment, which grows out of Catholic moral theology (O'Donnell 1956; Kelly 1958), separates obligatory treatments from those that are not obligatory on the basis of which treatment would benefit the patient. The distinction has been used extensively in philosophical and theological discussions and in some U.S. court decisions (President's Commission 1983).

Some people continue to use these terms to reflect the benefit of treatment for a patient. Others, including clinicians, often employ concepts about the aggressiveness of the technology in making distinctions between ordinary and extraordinary treatments. For example, more-aggressive treatments are more likely to be categorized as extraordinary, whereas others that are less aggressive are more likely to be categorized as ordinary, without regard for the benefit of the treatment for the patient. When the terms are used in discussions about the ethics of withholding treatment in particular cases, and in discussions about social policy, the assumption seems to be that the labels "extraordinary" and "ordinary" can be used to distinguish two mutually exclusive categories of treatments (for a critique of this position, see President's Commission 1983).

I observed that clinicians often used the terms as if treatments formed a continuum and were *ranked* as less or more extraordinary. Clinicians were consistent in the ranking of treatments in this way. There appeared to be much variation, hoewver, between clinicians in how extraordinary they considered a particular treatment to be in a particular case. This variation was associated both with conceptualizations of the benefit of the treatment for the baby and with the characteristics of the technology employed.

Survey on Neonatal Decision Making

The general clinical model and variations in the use of the terms "ordinary" and "extraordinary" are illustrated by responses to a survey I conducted on neonatal decision making. The survey was conducted in the spring of 1983, shortly after announcement of the Baby Doe directives. These regulations, issued by the Department of Health and Human Services, required treatment for infants born with handicapping conditions. They were promulgated following national publicity about an infant, referred to as Baby Doe, who was born with Down's syndrome and an intestinal defect. Without surgery, the infant could not be fed by mouth; the defect was surgically correctable. After a decision by the baby's parents and physician to withhold surgery and intravenous feedings and fluids, the infant died. (For more details, see Weir 1985; Lyon 1985; Levin 1985, Levin 1986.)

Results were based on questionnaires from 249 respondents. One hundred and thirty respondents attended a conference entitled "Which Babies Shall Live: Humanistic Dimensions of the Care of Imperiled Newborns," presented by the Hastings Center and Montefiore Medical Center and held in New York City; 97 respondents worked at the neonatal intensive-care unit of Columbia-Presbyterian Medical Center; and 22 of the respondents attended a retreat, also in New York, sponsored by the Columbia University Department of Obstetrics and Gynecology. Most of the respondents had professional experience working with catastrophically ill newborns. Forty-eight percent were nurses, 27 percent were physicians, and 18 percent were other health-care professionals. The remaining 7 percent included lawyers and journalists with a special interest in medical ethics. Although this is not a sample of a defined population, it does provide a set of responses from care givers most of whom are very knowledgeable about treatment decisions for catastrophically ill newborns.

The survey examined respondents' attitudes about choice and categorizations of treatments. The first portion of the questionnaire, based on

TABLE 6.1

Treatments Respondents Would Recommend
and Ratings on the Ordinary/Extraordinary Scale
for an Infant with Down's Syndrome (n=249)

Treatment	Would Recommend (percent)	Mean Score Ordinary/ Extraordinary
Intravenous feedings	91	1.3
Antibiotics	88	1.5
Surgery for intestinal defect	87	2.2
Cardiac catheterization	71	2.9
Open-heart surgery	59	3.7
Kidney dialysis	28	4.3

the work of Crane (1975), presented hypothetical vignettes of cases of newborns with critical conditions. For each case, respondents were presented with a list of treatments. The respondents were asked to assume that each treatment administered would increase the baby's chance of survival and that the parents' views were the same as their own. They were asked to indicate (1) if they thought it would be best to give or to withhold each of the treatments, and (2) how they would rate each of the treatments on a scale from one (ordinary) to five (extraordinary). For the purposes of the study, consensus was defined as agreement of at least 75 percent of the respondents; less than 75 percent was seen as indicating lack of consensus (Levin 1985).

Patterns of Treatment Recommendations

The first vignette, patterned after the case of Baby Doe, read as follows: "Baby 'A' is born with Down's Syndrome. Soon after birth, the baby is also found to have duodenal atresia, an intestinal defect which can be corrected by routine surgery. Without surgery, the baby cannot drink milk or other fluids by mouth." Most of the respondents said that they would recommend feedings, fluids, and surgery to repair the intestinal defect. There was no consensus about whether or not to recommend more-aggressive treatments. Fewer respondents said they would recommend open-heart surgery for this baby, and fewer still would recommend kidney dialysis if such treatments would be likely to prolong the life of the infant (see Table 6.1).

The treatment recommendations differed in the case of a baby with a more severe condition, presented in another vignette, but the pattern of

TABLE 6.2

Treatments Respondents Would Recommend and Ratings on the Ordinary/Extraordinary Scale for an Infant with Trisomy-13 (n=119)

Treatment	Would Recommend (percent)	Mean Score Ordinary/ Extraordinary
Before Chromosomal Analysis		
Nutrition and fluids	90	1.6
Antibiotics	81	1.9
Resuscitation	76	2.0
Respirator	65	2.4
After Chromosomal Analysis		
Nutrition and fluids	85	1.9
Antibiotics	60	2.8
Respirator	24	4.1
Surgery for cleft palate	14	4.0
Cardiac catheterization	13	4.5
Arrest page	10	4.6
Open-heart surgery	8	4.7

recommending less-aggressive treatments and withholding the more-aggressive ones remained the same. The vignette for that baby read as follows:

> Baby "C" was born with multiple congenital anomalies—low set ears, skin folds around the neck, a cleft palate, and cardiac anomalies—suggestive of Trisomy-13, a chromosomal anomaly that is always associated with severe mental retardation and severe physical impairments. Most of these babies die within the first few months and almost all die within the first year. If Baby "C" doesn't have Trisomy-13, he may have only correctable physical defects, or he may have uncorrectable physical and/or neurological defects.

After the description was interrupted for questions, the vignette continued: "Now suppose that after resuscitation in the delivery room Baby 'C' was breathing on his own and was admitted to the neonatal intensive care unit for evaluation. Two days later, chromosomal analysis indicated that he does indeed have Trisomy-13."

Table 6.2 presents the treatment recommendations and the mean scores on the ordinary/extraordinary scale for the baby with Trisomy-13. As in the case of the infant with Down's syndrome, more respondents recommended giving the less-aggressive treatments and withholding the more-aggressive ones. This was true both before and after a definitive diagnosis was made. Before chromosomal analysis, there was consensus to give nutrition and fluids, antibiotics, and resuscitation, but there was no consensus about putting the baby on a respirator.

After diagnosis of Trisomy-13, there was consensus to withhold open-heart surgery or use of a respirator; there was no consensus about antibiotics. It is interesting to note that there was agreement to give nutrition and fluids, the least-aggressive treatment modality required. Differences between treatments selected in the two cases show that differentiations are made based on both characteristics of patients and characteristics of treatments.

Respondents recommended more treatments for the infant with Down's syndrome, who could be expected to have a better quality of life than the baby with Trisomy-13. In each case, respondents were more likely to recommend giving the less-aggressive treatments and were more likely to recommend withholding more-aggressive treatments.

The ranking of treatment choices according to the aggressiveness of treatment was reflected not only in variation in the treatment choices between cases but was also reflected in the pattern of individual respondents' recommendations for treatment. If a particular respondent recommended withholding a particular treatment in one case, that respondent was likely to withhold all other more-aggressive treatments for that case as well. And conversely, if a respondent recommended giving a particular treatment, that respondent was likely to recommend giving all other less-aggressive treatments in that case.

The extent to which the pattern of treatment recommendation of individual respondents reflected the same ordering of treatments as the respondents as a group was measured by use of Guttman scale analysis. The coefficient of reproducibililty for the treatment recommendations for each case exceeded 0.92, which indicates that for each of the cases, the respondents' answers were highly correlated with the ordering of treatments in terms of aggressiveness.

Differences in treatment recommendations according to background characteristics of respondents were investigated using an "overall aggressiveness score" based on responses about treatment in a number of cases. Significant differences ($p<0.05$ on a one-tailed t-test) were found between members of some professional and religious groups, although there was much individual variation. Catholic respondents, in general, and Catholic members of the same occupational group, in particular, recommended significantly more aggressive treatment than did Protestants, whereas the average score of Jewish respondents was intermediate. The average overall aggressiveness score of neonatal nurses was significantly higher than that of neonatologists. The pattern of recommendations of members of each professional and religious group corresponded to the overall pattern of giving less-aggressive treatments while withholding more-aggressive ones.

Although there were significant differences in treatment recommendations when occupational and religious categories were compared at the group level, occupation and religion were not significantly associated with the overall aggressiveness score after adjustment by multiple regression, a statistical technique used to explore the separate effects of age, sex, professional experience, religiosity, religion, and occupation. The entire set of variables explained only 15 percent of the total variation in aggressiveness scores. Only age was a statistically significant explanatory factor ($p<0.05$), accounting for approximately 10 percent of the variation; older respondents were less aggressive in their treatment recommendations than were younger respondents.

Among respondents to this questionnaire, being a nurse, being Catholic, and being young were all associated with each other, and all tended to be associated with recommending more-aggressive treatment. However, individual doctors and nurses and Catholics and non-Catholics who were the same age did not differ significantly in their treatment recommendations. Indeed, many of the older, more-experienced nurses recommended withholding treatment more than many of the younger residents and neonatal fellows.

Ranking of Ordinary and Extraordinary

In addition to illustrating patterns in recommendations for giving or withholding treatment, the results of the questionnaire illustrate patterns in the way that respondents categorized treatments as ordinary or extraordinary. The influence of both patient and treatment characteristics may be seen in the responses of the clinicians to questions about the categorization of treatments on a five-point scale from one (most ordinary) to five (most extraordinary). Whereas there was a high degree of consistency in the *relative ranking* of treatments, there was much variation in the *labeling* of treatments as ordinary or extraordinary, from case to case. For example, almost all respondents ranked cardiac surgery as more extraordinary than intestinal surgery but less extraordinary than kidney dialysis. The range of respondents' categorizations, however, spanned the whole continuum; that is, for each treatment, some of the respondents indicated "1" for the most ordinary, and others indicated "5" for the most extraordinary.

The findings shown in Table 6.3 indicate that respondents consistently withheld more treatments and labeled treatments as more extraordinary for a more seriously affected case. The baby with Down's syndrome was given more treatments, and treatments were labeled as less-extraordinary, compared to the baby with Trisomy-13.

TABLE 6.3

Comparison of Ratings on the Ordinary/Extraordinary Scale by Patient Condition and Treatment[a]

	Ratings				
	Ordinary--------Extraordinary				
Treatment and Patient Conditions	1	2	3	4	5
IV feedings/nutrition and fluids:					
Down's syndrome and duodenal atresia	83	8	5	2	2
Trisomy-13 with cleft palate	63	16	4	7	10
Surgery for:					
Duodenal atresia (Down's)	33	33	22	7	5
Cleft palate (Trisomy-13)	4	9	18	22	47
Open-heart surgery for a:					
Baby with Down's syndrome	4	9	27	27	33
Baby with Trisomy-13	2	0	6	11	81

[a]Numbers represent the percent of respondents who would assign each treatment a particular rating on the ordinary/extraordinary scale. For treatments for the baby with Down's syndrome, n=249; for treatments for the baby with Trisomy-13, n=119.

Respondents who recommended a treatment were more likely to categorize a treatment as more-ordinary than were those who would not recommend the same treatment. The treatment recommendations and categorizations of treatment for the infant with Down's syndrome are reported in Table 6.4. Respondents, however, did not base categorizations as extraordinary or ordinary on presumed benefit of treatment alone. Categorizations also depended on the characteristics of the technology. Most respondents, including those who would withhold intestinal surgery from a baby with Down's syndrome, did not rate such surgery as very extraordinary. On the other hand, most respondents, including those who would recommend kidney dialysis, rated such treatment as fairly extraordinary.

These findings suggest that respondents base treatment recommendations and categorizations of treatments as ordinary and extraordinary not only on the characteristics of the patients but also on the characteristics of the treatments. The finding that respondents would recommend more treatments for an infant with Down's syndrome than for a baby with Trisomy-13 indicates that they do consider the characteristics of the patient in making treatment recommendations and categorizations. However, the fact that for both infants respondents would recommend giving less-aggressive treatments while withholding more-aggressive ones indicates that they also consider the aggressiveness of the treatment. The correlation of recommendations

TABLE 6.4

Ratings on the Ordinary/Extraordinary Scale by Recommendations Concerning Treatment for the Baby with Down's Syndrome[a]

Treatment Recommendations		Ratings: Ordinary-------Extraordinary 1	2	3	4	5
Intestinal surgery						
Would recommend	n=216	37	36	19	6	2
Would not recommend	n=31	7	10	45	16	23
Kidney dialysis						
Would recommend	n=67	10	15	21	22	31
Would not recommend	n=173	1	3	6	16	74

[a]Numbers represent the percent of respondents within each treatment category (recommending or not recommending treatment) who would assign the treatment a particular rating on the ordinary/extraordinary scale.

to give or withhold treatments based on the condition of the baby and categorizations of treatments as ordinary and extraordinary indicates that there is an association between categorizations and the presumed benefit of treatment. Categorization of treatments as ordinary and extraordinary, however, also depends on characterizations of treatments. Sometimes clinicians categorize treatments that they would recommend withholding as ordinary and also categorize treatments they would recommend giving as extraordinary.

Questions About the Use of Technological Criteria

Bioethicists, clinicians, parents, and other members of society need to address questions about what criteria are relevant in guiding treatment choice. One might argue that only patient characteristics—for example, the future quality of life of the baby—should be relevant in making decisions about care. Why then are technological criteria considered in making decisions about the care of newborns? Are technological criteria relevant for making appropriate treatment choices?

There are at least two reasons why technological criteria are considered. First, in all situations of medical decision making, clinicians must choose what treatments they will recommend. The aggressiveness of treatments based on such characteristics as invasiveness and risk, physical and mental pain, and cost/benefit ratios are relevant in making evaluations of treatment choices. In general, clinicians try to treat patients using the least-aggressive treatments that will preserve life and promote health. Even when evaluation of patient characteristics leads clinicians to conclude that preserving a patients' life would not be of benefit, they must still act; they still face decisions about whether specific treatments should be given or withheld. Continuing to use technological criteria allows clinicians to make decisions about excluding treatment for these infants in the same manner they use to make decisions about the care of other infants.

Second, in making decisions using technological criteria, clinicians are able to conceptualize their choices as decisions about treatments rather than as decisions about whether a particular baby should live or die. There are strong ethical traditions that place value on all human life. There is a fear of committing immoral acts, such as those associated with the atrocities of Nazi euthanasia that were practiced on Jews, homosexuals, the disabled, and other people who were seen as undesirable and were "selected for death." This fear leads to a strong norm against practicing active euthanasia. As a result, decisions are sometimes made to give an intermediate level of treatment that may allow an infant to either die or survive. In such situations, clinicians can conceptualize that the outcome is determined by the baby, by nature, or by God rather than by their decision.

Are technological criteria morally relevant for making appropriate treatment choices? There may be some situations in which treatment characteristics are relevant in the decision-making process; in other cases, such considerations may preclude making appropriate decisions. Treatment characteristics may be relevant in situations in which treatments may cause unacceptable pain and suffering. For example, repeated painful surgeries for a baby with little chance for survival could be seen as inhumane treatment.

Attention to technological criteria may also be relevant in distinguishing between acts that cause death—active euthanasia—and acts that allow death to take place—passive euthanasia. For example, preventing a child from breathing or withholding bottle or breast feedings that could be used to nourish a normal child may be seen as active euthanasia; refraining from using such sophisticated medical technology as respirators and intravenous feeding to treat a severely ill newborn could be seen as passive euthanasia. Such distinctions may be of value in differentiating treatments that would provide appropriate care from others that would prolong the dying process.

In other situations, however, consideration of characteristics of treatments may divert attention from the morally relevant criteria. If, on the basis of patient characteristics, parents and clinicians decide that treatment to prolong life is not in the best interests of a severely compromised infant, then the characteristics of the treatments themselves may not be relevant in the decision-making process. For a comatose infant with no hope of ever being able to recognize caretakers or derive pleasure from other aspects of the environment, neither "extraordinary" treatments such as the use of a respirator nor "ordinary" treatments such as giving antibiotics may be of benefit.

Decision making about the care of catastrophically ill newborns raises many ethical questions. Parents, clinicians, bioethicists, social policy makers, and others face difficult choices concerning how to protect the best interests of such babies. There is need to identify the criteria governing current treatment choices and to evaluate critically the moral relevance of the criteria employed. The use of categorizations based on technological criteria, such as the categorization of treatments as ordinary or extraordinary, in making decisions about care is one of the issues in need of further examination.

Acknowledgments

Support for some of the research presented in this paper came from the Project on Issues of Values and Ethics in Health Care, College of Physicians and Surgeons, Columbia University and from a Charlotte W. Newcombe, Woodrow Wilson Dissertation Fellowship. I wish to thank Dr. John Driscoll and the staff of the Columbia-Presbyterian newborn intensive-care unit for their assistance in conducting research and Drs. Arthur Caplan, John Colombotos, Ann Dill, Tom Murray, and Kathy Powderly for their valuable comments on earlier versions of this paper.

References

Aries, Phillippe. 1982. The Hour of Our Death. New York: Alfred Knopf.

Capron, Alexander, and Eric J. Cassell. 1984. Care of the Dying: Withholding Nutrition. Hastings Center Report 14:32–38.

Crane, Diana. 1975. The Sanctity of Social Life: Physicians' Treatment of Critically Ill Patients. New York: Russell Sage Foundation.

Guillemin, Jeanne Harley, and Lynda Lytle Holmstrom. 1986. Mixed Blessings: Intensive Care for Newborns. New York: Oxford University Press.

Kelly, Gerald. 1958. Medical-Moral Problems. St. Louis: Catholic Hospital Association of the United States and Canada.

Levin, Betty Wolder. 1985. Consensus and Controversy in the Treatment of Catastrophically Ill Newborns: Report of a Survey on Which Babies Shall Live. Humanistic Dimensions of the Care of Imperiled Newborns, Thomas H. Murray and Arthur L. Caplan, eds., pp. 169–205. Clifton, New Jersey: Humana Press.

______. 1986. Caring Choices: Decision Making About Treatment for Catastrophically Ill Newborns. Ph.D. thesis, Columbia University.

______. 1988. Technological, Social and Ideological Factors in Care and Decision Making for Catastrophically Ill Newborns: The Cultural Context of Baby Jane Doe. Childbirth in America: Anthropological Perspectives, Karen M. Michaelson, ed. South Hadley, Massachusetts: Bergin and Garvey.

Lyon, Jeff. 1985. Playing God in the Nursery. New York: W. W. Norton and Company.

O'Donnell, Thomas J. 1956. Morals in Medicine. Westminster, Maryland: The Newman Press.

Parsons, Talcott, Renee C. Fox, and Victor M. Lidz. 1972. The "Gift of Life" and Its Reciprocation. Social Research 39:367–415.

President's Commission for the Study of Ethical Problems in Medicine and Biomedical and Behavioral Research. 1981. Defining Death: Medical, Legal, and Ethical Issues in the Determination of Death. Washington, D.C.: U.S. Government Printing Office.

———. 1983. Decisions to Forego Life-Sustaining Treatment: Ethical, Medical, and Legal Issues in Treatment Decisions. Washington, D.C.: U.S. Government Printing Office.

Reich, W. T., ed. 1978. The Encyclopedia of Bioethics, vol. 2, pp. 717–751. New York: The Free Press.

Reiser, Joel, Arthur Dyck, and William Curran. 1977. Ethics in Medicine. Cambridge, Massachusetts: MIT Press.

Silverman, William A. 1980. Retrolental Fibroplasia: A Modern Parable. New York: Grune and Stratton.

Weir, Robert. 1985. Selective Nontreatment of Handicapped Newborns. New York: Oxford University Press.

7

From Principles to Practice: Life-and-Death Decisions in the Intensive-Care Nursery

Renee R. Anspach

This chapter returns to a classical problem in a new context: the relationship between theory and practice, ideas and action. The ideas are those of bioethicists; the practice is that of physicians and other clinicians; and the setting is intensive-care units for newborn infants. Without plumbing the depths of the is/ought controversy, I assume at the outset that the ideas of bioethicists may be made relevant to the realities of clinical practice, and my purpose in this chapter is to suggest ways in which theory and practice can be brought into better accord.

The setting—neonatal intensive-care—is an emerging specialization in medicine and nursing that has undergone rapid development since the 1960s. Conditions associated with prematurity are the most common cause for hospitalization in the intensive-care nursery, followed by patients with congenital anomalies, that require immediate treatment (Budetti et al. 1980). Available epidemiological data suggest that the advances of newborn intensive care have contributed to a reduction in mortality among critically ill newborn infants, although the effect of neonatal intensive care upon morbidity remains the object of some controversy (Budetti et al. 1980). However, in the face of these therapeutic triumphs, it is at the same time also acknowledged that advances in medical technology and surgical intervention have made it possible to sustain the lives of infants whose long-term prospects for survival may be limited or of infants who may survive with serious physical or mental disabilities. Consequently, several physicians, nurses, and parents have reached decisions that not all infants are to be treated actively. Based on the few published reports in medical journals that have appeared over the years, there is reason to believe that these life-and-death decisions have been made for some time in intensive-care nurseries throughout the country (Duff and Campbell 1973).

As advisory commissions and medical practitioners confront dilemmas for which the canons of medical training provide few guidelines, increasingly they look to experts for guidance. Bioethicists are one group of experts who have assumed an influential role. When I refer to *bioethics,* I am using the term quite broadly to refer to any attempt—whether by a physician, a philosopher, or a theologian—to examine the ethical principles upon which decisions in medicine or the biological sciences are or should be premised. In other words, I am using the term to refer to a set of aims and a characteristic discourse rather than to a particular discipline or disciplines.

Bioethicists have addressed a number of issues, two of which are particularly salient. The first is whether, and under what circumstances, it is appropriate to withhold life-sustaining treatment from critically ill newborn infants. Most controversial are those decisions that concern withholding treatment from infants on the basis of their anticipated "quality of life." At one end of the continuum of opinion are those writers, most notable among them Paul Ramsey (1978), who argue that quality-of-life decisions, in which treatment is withheld from seriously disabled infants, erode the fundamental equality of human beings. Instead, Ramsey proposes a "medical indications policy," which justifies withholding lifesaving treatment only from infants who are already dying. Other writers, such as Tooley (1972), Lorber (1975), and Engelhardt (1975), suggest that it is justifiable to withhold life-sustaining treatment from infants with serious physical and/or mental disabilities who may impose undue hardships on their families and society. Midway between these positions are those bioethical writers who contend that quality-of-life decisions may be ethically justified in limited circumstances—as, for example, when an infant is likely to live with intractable pain or when infants are unable to participate in human relationships (Jonsen and Garland 1976; McCormick 1974).

Another highly debated issue concerns who should have the authority for making life-and-death decisions. On this point, ethical opinion is again sharply divided. Some writers have suggested that parents, with the assistance of their professional advisors, should have the final authority in decision making (Shaw 1973); others have suggested that hospital boards should develop policies to guide decision making (Waldman 1976); another view is that the state ("society") should regulate decisions to terminate care, through legislative bodies or government commissions (Robertson 1977). In short, those who seek the counsel of experts are likely to encounter a considerable diversity of opinion.

Bioethicists are not only divided in their views, but, moreover, they have diverse visions of their own projects and professional role. The more academically oriented philosophers view bioethical analysis as a theoretical enterprise and use decisions in medicine or biology as vehicles for examining classical issues in moral philosophy. Others use philosophical skills to clarify the concepts, principles, and precepts implicit in medical decisions. Still others have a more "applied" vision of the enterprise; following quasi-consultant roles, they attempt to develop principles and policies that can assist clinicians in reaching life-and-death decisions.

The "applied" emphasis has assumed increasing importance as bioethicists are requested to serve as consultants to intensive-care nurseries, to publish guidelines for decision making in medical and scientific journals, to develop courses that are integrated into medical school curricula, and to serve on policymaking commissions. In fact, Baby Doe directives have stipulated that each hospital appoint ethics committees, thereby giving quasi-legal recognition to the bioethicists who serve on them. The ventures of bioethicists into the applied realm have been met with a mixed reception by medical practitioners. Some clinicians have welcomed the bioethicist's presence; some have viewed the bioethicist as lending legitimacy to decisions they would make in the first place; and others, among them physicians and ethicists, have been more resistant, criticizing much of bioethical analysis for its lack of relevance to clinical practice (Lo and Jonsen 1981).

The source of this resistance is the subject of this chapter, in which I examine some of the problems in applying bioethical analysis to actual life-and-death decisions in the intensive-care nursery. My arguments are grounded in data collected during sixteen months of fieldwork in two intensive-care nurseries, which differed with respect to size, prestige, referral patterns, and the demographic composition of their clientele. One setting, a twenty-two-bed nursery in an elite institution, functioned as a major referral center and served a demographically heterogeneous clientele. The other, a much larger (sixty-bed) nursery, did not function as a referral center and served a largely indigent clientele.

I will discuss four problems frequently not addressed in bioethical analysis: (1) uncertainty in medical decision making; (2) negotiations between parents and practitioners; (3) negotiations among staff; and (4) the social context of decision making. These considerations, which often elude bioethical analysis of life-and-death decisions, are, in fact, very fundamental concerns for those who make them. My discussion will conclude with some suggestions as to how bioethicists can expand their analysis so as to embrace the actualities of medical practice.

Uncertainty in Medical Decision Making

A first issue concerns the many uncertainties that attend life-and-death decisions in the intensive-care nursery. Because their major concern is with ethical dilemmas, most bioethicists have emphasized those medical conditions in which these dilemmas are posed with crystalline clarity. For example,

Down's syndrome has received much attention in the bioethics literature because it poses a significant ethical dilemma with particular acuity: whether lifesaving treatment may be withheld from infants whose major defect is subnormal intelligence. Other frequently discussed but statistically rare conditions are spina bifida, Tay-Sachs disease, and anencephaly. To be sure, some of the literature has emerged in the wake of legal and public controversies surrounding the Baby Doe cases, which concerned infants with Down's syndrome and spina bifida. However, even before these controversies emerged, most ethical discussion focused on congenital anomalies in which the prognosis of those affected is, within a certain range, predictable. But, as Jonsen and Lister have noted (1978), many newborns treated in the nursery have a prognosis that is uncertain. Among the subjects of my research, more than one-third of life-and-death decisions involved infants whose medical futures were considered uncertain or were the subject of considerable debate among providers. These cases included premature infants thought to be on the frontiers of viability, infants having atypical clinical courses, infants with birth asphyxia or brain hemorrhages whose future mental potential was viewed as difficult to assess, and infants who had received highly innovative medical or surgical treatments.

Although the prognosis of any patient may arguably be said to be uncertain, as discussed in Chapter 4, the highly innovative nature of newborn intensive care makes it difficult to predict the viability or intellectual potential of many critically ill newborn infants. In fact, even when practitioners could agree on the ethical *principles* in a life-and-death decision, often they could not agree on the infant's *prognosis*, as the case of Robin Simpson, presented in Chapter 4, suggests.

Perplexing prognoses introduce additional dilemmas into the decision-making process, for they demand that physicians consider what level of certitude is required to reach a life-and-death decision and by what standards relative certitude should be established. These are dilemmas that elude many bioethical treatments of decision making. Bioethicists have confined their analyses to cases that are ethically complex but prognostically simple, but—as has been suggested by the many cases actually encountered in the nursery—most decisions clinicians must make are both ethically and prognostically complex.

Negotiations Between Parents and Practitioners

Ethical opinion is somewhat divided as to whether, and to what extent, parents should be incorporated into the decision-making process. Some writers (Ramsey 1978) have argued that infants have rights apart from those of the parents, and, for this reason, decisions should be made by physicians in accordance with the dictates of ethical policy. Some (Zachary 1978) have suggested that physicians, owing to their superior knowledge and expertise, should be the principal decision-makers. Others, notably Shaw (1973) and Duff and Campbell (1973), have advocated a policy of informed consent. In cases of infants who will have limited prospects for long-term survival, or who will survive with serious physical and/or mental disabilities, these authors suggested, parents have the right to decide whether their infants should live or die. My arguments are directed to those bioethical writers who advocate policies of informed consent, for they leave unanswered a basic question—namely, *how* that consent process is to take place. For simply to articulate the principle that parents should be incorporated into the decision-making process does not consider the highly complex and delicate negotiations that often take place between parents and providers.

Consider, for example, a critical issue: How should diagnostic and prognostic information be presented to parents? Should parents be presented with a *range* of prognostic outcomes, or should prognostic information be weighted toward the low or the high end of a continuum? Should parents be presented with the *option* of discontinuing life support, or should they be provided with a recommendation by the staff? More to the point, should parents be treated as the principal participants in the life-and-death decision (informed consent) or should they be viewed as giving *assent* to a decision already made by staff? The quotations following, taken from interviews with physicians and observations of staff conferences, illustrate the contrast between informed consent and assent in actual practice:

Assent Model	Consent Model
(Infant with Down's syndrome)	(Infant with Down's syndrome)
All children with Down's syndrome are mentally retarded, and many develop multiple medical problems. Your baby will always have to have a colostomy. Do you understand what that means? Many children with Down's syndrome are able to lead happy, productive lives. . . .	Some children with Down's syndrome are able to lead happy, productive lives, and often are able to work in sheltered workshops; others are not able to function as well. Some parents find that the experience of caring for a child with Down's syndrome can be rewarding; others find it more stressful. Would you like to meet some parents who have had similar experiences? [hypothetical example]

(Dying infant)	(Dying infant)
We have had a meeting and consulted with several specialists. . . . All of us agree that we have tried everything, and there is nothing more that can be done. Now the medicines are keeping the baby alive, and we would suggest taking the baby off the machines to let you hold him. Keeping the baby alive will only prolong his suffering.	________'s kidneys are starting to fail. We've talked to a number of specialists and we all agree that there is nothing more that can be done. ________ will die in a matter of days. We have two options. One is to continue to give ________ the medicines that are keeping him alive. The other is to stop giving the medicine, and let you hold the baby. Would you like to think about it?

As medical sociologists have argued since Parsons (1951), there is an ineluctable "competence gap" between doctor and patient, and, for this reason, physicians hold hegemony over the decision-making process. However, as the previous examples suggest, even within these constraints physicians may structure the conversation so as to allow considerable latitude for patient participation. If a model of informed consent is practiced, there are a number of ways in which the role of the parents in life-and-death decisions can be expanded. Clinicians may facilitate parent involvement by posing open-ended questions, by presenting parents with a range of prognostic outcomes, by introducing parents to others in similar situations, and by posing options for decision making. If, on the other hand, parents are seen as providing *assent* to a decision already made by staff, there are a number of ways in which providers may, intentionally or unwittingly, frame or shape the parents' decisions. These may include: presenting prognostic outcomes leaning toward the positive or negative end of the continuum, consulting parents only *after* the staff had already reached its decision, presenting a recommendation unacceptable to the parents, or articulating a moral precept with which parents would find it hard to disagree—such as, "Keeping the baby alive will only prolong his [or her] suffering."

In both intensive-care nurseries that were studied, physicians acknowledged the importance of "incorporating parents into the decision-making process." However, as most physicians acknowledged that they virtually controlled the decision-making process, the model of assent was the one most frequently used. More often than not, parents were asked to agree to decisions already reached by staff. However, given the large number of parents and providers who were involved in the decision-making process, there was considerable variation in the extent to which parents were encouraged to participate. The degree to which parents actively participated in the decision-making process varied according to a number of contingencies, including: how staff members conceived their own role; staff perceptions of the educational attainment or "medical sophistication" of the parents; staff perceptions of parents' religious beliefs (staff members were more reluctant to discuss life-and-death decisions openly with parents viewed as religious Catholics); the degree to which staff felt convinced that its decisions were correct (when staff members were ambivalent, they gave greater latitude to the parents to make the decision); and staff perceptions of the possible psychological consequences for the parents that might ensue from incorporating them into the decision-making process.

A frequently cited "folk" psychological theory was that parents who were asked to play an active role in life-and-death decisions would later experience guilt. As one neonatologist stated, "I don't believe you should ask the parents outright since that lays a guilt trip on the parents." This belief, however, has little support in the literature concerning the social psychology of decision making. Janis and Mann (1977) have noted that "post-decisional regret," or guilt, is less likely to ensue when decisions are made through what the authors call "vigilant information processing." In order for vigilant information processing to occur, participants in consequential decisions should participate actively in decisions and be given time to weigh the alternatives carefully. Conversely, post-decisional regret, or guilt, is more likely to ensue when participants are not provided with information or when they are given insufficient time to weigh the alternatives. These considerations suggest that a well-intentioned but paternalistic attempt to protect parents from guilt may, ironically, produce the very effects it is designed to minimize.

Additional issues complicated the consent-process that arose on more than one occasion in both nurseries that were studied. Parents may selectively attend to and comprehend the information they receive (Cicourel 1983). Educated parents with greater medical knowledge may have more latitude in negotiation with staff. A nursery frequently delegates responsibility for communicating with parents to a number of providers. Because providers may hold differing opinions about a life-and-death decision, parents may be confused by conflicting information. Interpreters may introduce complications into conferences with parents by unintentionally reframing what has been said. When the nursery is a referral center and patients are transported from outlying areas, one or more parents may be separated from the infant during critical portions of the life-and-death decision. It is ironic that the mother, who may remain hospitalized while the infant is transported to a major referral center, may be excluded from decision making but is often the one who bears most of the consequences for subsequent decisions that are made.

Finally, the parents are not, as is sometimes suggested in the ethical literature, a single unit. One parent may dominate the interaction in decision-making conferences, and/or there may be disagreement between the parents as to whether their baby should live or die. The following case illustrates the tragic consequences that can ensue when parents disagree and one parent is allowed to prevail:

> Joan Gomez and Robert Smith gave birth to 800-gram (1-pound, 11-ounce) twins, born fourteen weeks prematurely. The couple was unmarried, on welfare, and lived in a rural area more than fifty miles from the nursery that was being studied. One week after birth, one of the twins developed a very serious cerebral hemorrhage. The parents were called into a conference to discuss termination of life support. They were told that the baby had suffered a serious brain hemorrhage and that it was quite likely that he would be seriously retarded, probably unable to walk or talk. At one point in the conversation Ms. Gomez said that she felt it would be better if the baby died. The father, however, dominated the interaction. He adamantly insisted that the staff try to keep the baby alive. He told the staff that he had a nephew who suffered a brain hemorrhage and became a perfectly normal child. The staff continued to treat the baby. In the ensuing months the parents visited the infant only once. They ultimately abandoned the babies, who are now in foster homes.

This case illustrates a number of serious ethical and social dilemmas that can arise from disagreement between the parents. When parents disagree about the fate of their child, whose wishes should prevail? Should staff have acceded to the wishes of the father, who seemed more vocal and adamant? Or should more weight have been given to the opinion of the mother, who, in all likelihood, would bear the social and economic consequences of caring for a seriously disabled infant? The case reflects other organizational problems that occur in referral centers in which parents are separated from their infants by significant geographical distance. What is important for our purposes is the fact that parents may approach the life-and-death decision with very different values and interests and may have different conversational resources with which to promulgate their views. At minimum, it would seem that staff are obligated to listen to and carefully elicit the opinions of both parents.

In calling attention to the more problematic features of parent-staff communication, I do not mean to suggest that staff are inept. Quite the contrary. In both nurseries that were studied, most conferences with the parents were conducted with sensitivity and consideration. My point is simply that it is difficult to implement the principle of informed consent without understanding the process by which the parents are informed and their consent obtained.

Negotiations Among Staff

Bioethicists, then, have directed their attention to the principles at stake in life-and-death decisions rather than to the process by which these decisions are reached. This limitation stems in part from the decision of some bioethicists to focus their analysis on a single decision-maker—usually the physician; less often, the nurse. This focus implies that the actor reaches decisions alone, apart from the influences of others. The life-and-death decision is a complex social activity that embraces several participants: attending neonatologists, consultants, residents, fellows, nurses, and parents. Each set of participants may have conflicting conceptions of the "facts" brought to bear on the decision, as well as conflicting conceptions of the "right" thing to do. Although attending neonatologists are superordinate in the medical hierarchy and may have the greatest authority in decision making, in recent years there has been an increasing attempt in many nurseries throughout the United States to incorporate subordinates into the decision-making process. For this reason, medical rounds, case conferences, and ethics conferences have become arenas in which differing and often conflicting points of view are negotiated in a process of reciprocal influence. To this extent, decisions have come to reside not with a single participant but rather in the interactions between participants.

The participants do not, however, negotiate equally. Nurses and residents who disagree with attending neonatologists are free to defy them but face possible repercussions to their careers. As must be evident, attending neonatologists must face the consequences of reaching decisions that are at odds with hospital policy and the opinions of colleagues. To the extent that participants in life-and-death decisions take these considerations into account, it is possible to speak of organizational constraints on decision making.

The Social Context of Decision Making

Within the Hospital and Nursery

Much of the literature on the ethics of neonatal intensive care individualizes the decision-making process, obscuring not only the way in which decisions are reached but the social context in which they take place. A major feature of that context is the organization of the intensive-care nursery as a work environment. Complex organizations in general, and the newborn intensive-care unit in particular, exert several significant influences on decision making.

First, hospitals as "advisory bureaucracies" (Goss 1961) allocate different resources to participants in life-and-death decisions, enabling them to promulgate their respective points of view; this differential creates probabilities that the views of those favored will prevail. Medical knowledge is one obvious resource. But other resources include the relative authority and autonomy of occupational groups in the medical division of labor and the relative ability of these groups to impose sanctions (Freidson 1970).

Second, organizations structure the possible consequences that ensue from engaging in a given course of action. Because decisions take place within a framework of power relationships, health professionals who consider reaching decisions that are at variance with the wishes of influential colleagues and superiors must contend with the potential consequences of these decisions. Although these consequences do not entirely preclude certain options or determine completely the way in which individuals will act, they do influence the probabilities that a certain line of action will be chosen.

Third, organizations create commonalities of interest and generate subcultures. Each occupational group in the intensive-care nursery confronts a common set of problems created by the daily work environment; to the extent that members within each group communicate informally with one another, a common perspective on decision making frequently evolves. In a larger research project (Anspach n.d.), I examined the decisions of three groups. The attending neonatologists were academicians as well as clinicians, and they often experienced a conflict between research and clinical roles. The interns and residents, who rotate through the nursery on a monthly basis, viewed themselves as overworked and were rewarded for their mastery of technical knowledge. Nurses sustained a relatively continuous contact with both parents and infants. Each set of participants had a set of interests transcending, or even conflicting with, patient care. Attending neonatologists had an interest in advancing research; residents had an interest in learning and in minimizing uninteresting "scut" work (Becker 1961); and nurses had an interest in caring for acutely ill infants who are both medically and socially responsive.

These conflicting interests are sometimes revealed when medical treatments are applied involving surgical interventions that are viewed as experimental. Conflicts over the appropriate use of innovative or unusual procedures frequently ensue. The following case illustrates this conflict:

> Roberta Z. was the ninth child of a forty-four-year-old mother. Roberta's parents were farm workers in a valley situated several miles from the nursery. Shortly after birth, Roberta was diagnosed as having an unusual lesion, an arteriovenous malformation in the vein of Galen (an abnormal connection between a vein and an artery in the brain). The doctors at the referring nursery first told the parents that the lesion was inoperable. However, Roberta was transported to the nursery, where a neurosurgeon had pioneered a procedure for this lesion

and had operated successfully on adults and older children. This operation, which had never before been attempted on an infant, consisted of inserting fifty feet of fine wire into the malformation in order to induce a blood clot that would close off the lesion. Altogether, four operations were performed on Roberta: two to insert the fine wire and two more to clip off the blood vessels that fed into the malformation.

After the third surgery, the nurses pushed repeatedly for a conference to decide whether treatment should be discontinued. The specialists, neonatologists, fellows, and most residents felt that continued surgery was justified, but the nurses became progressively insistent that the fourth operation should not be performed, and many commented that Roberta had "become an experimental animal for the neurosurgeons." As one nurse said,

> She was an experimental animal for the neurosurgeons. They had no feeling for her suffering as a person since they don't sit here and watch her day and night. One night she was having convulsions and her head was leaking cerebrospinal fluid all over, so we call the neurosurgeon. He came in and bandaged her head and I mean he scrubbed her head with betadyne. Now I grant you she had an infection but it was the way he did, like you would scrub a table, and then we said, "Why don't you give her some xylocaine?" and he said, "Oh no," and stitched up her head, without anesthesia. This is what made us angry; she's not a person to them, she's just their experiment.

After the fourth operation, it was agreed by the staff in a conference that no further operations were to be performed. The parents were told that Roberta would die and that nothing further could be done. She lived for another month, finally dying of a meningitis that was not treated.

The case raises a number of complex issues that have been addressed elsewhere in this chapter—for example, the difficulty in gleaning the consent of parents separated from the staff by invisible lines of language and social class. What is significant at this point is that the conflict between the nurses and the medical staff represents different resolutions of the "experiment/therapy" dilemma. According to Fox and Swazey (1974), research physicians frequently are forced to strike a delicate balance between an eagerness to advance scientific knowledge and a commitment to the patient who is being treated. What is perhaps at issue is differing conceptions of "success." For the physicians interviewed after Roberta had died, the operations were not a failure, as they held the possibility of advancing the care of future patients. For the nurses, however, the surgeries were far from successful, because the nurses' commitment to Roberta outweighed a concern with advancing knowledge and improving the care of future patients.

The case also reflects the differing social positions and work experiences of the participants. Nurses, who observe and care for patients on a daily basis, not infrequently accuse research physicians of sacrificing matters of the heart on the altar of scientific zeal. This is particularly apt to be the case when the principal participants are specialists, outsiders to the nursery. In the view of the nurses, these practitioners are insulated from the concrete consequences of their actions. This case, then, reflects a common precept concerning the sociology of knowledge: that conflicting values and interests arise out of differing social and organizational positions of the participants.

There is a fourth, and more subtle, way in which organizations structure the nature of life-and-death decision making. Organizations vary in their capacity to allocate qualitatively different types of information to those who make life-and-death decisions. Elsewhere (Chapter 4 in this book), I have argued that members of the nursery staff, by virtue of their very different daily experiences in the intensive-care nursery, may come to develop conflicting conceptions of the very "facts," "data," and "evidence" to be used in reaching life-and-death decisions.

Public Policy

Life-and-death decisions are also influenced by developments outside the walls of the intensive-care nursery. Shifting legal rulings, articles in the mass media, and public policy pressures toward cost containment—to name a few of the more obvious developments—conspire to create a wider climate in which decisions must be reached. Less obvious, but even more significant, is the impact of public policy decisions affecting resource allocation outside the nursery upon life-and-death decisions. It has been society's decision to lower infant mortality rates by allocating resources to the technological interventions of neonatal intensive care. Society has chosen not to allocate major resources to preventive strategies and improved access to prenatal care and nutrition, which would lower the incidence of prematurity and birth defects. It is entirely possible that public health alternatives might require other changes, such as income redistribution, increased regulation of occupational safety and industrial pollution, and major changes in the distribution of health care—hardly likely to occur in the contemporary political climate.

But the technological solution is the one that has been chosen, and it has been an expensive choice. Neonatal intensive care ranks among the most costly of therapeutic modalities. Expensive equipment, high nurse-patient ratios, and third-party reimbursement practices (which encourage the use of more expensive care) all conspire to create costs that strain the imagination. A leading study, undertaken in 1980, estimated national annual expenditures on newborn intensive care to be about $1.6 billion, or an average of $8,000 per patient (Budetti et al. 1980). Another group of researchers, comparing expenditures among patients with high-cost hospitalization ($4,000 in a year), found newborn cases to be among the most expensive, averaging $20,000 per patient and thereby exceeding costs of two of the most expensive adult services, for neoplastic and for circulatory disease.

What of the effect of neonatal intensive care on mortality and morbidity? In order to evaluate the epidemiologic impact of neonatal intensive care, Budetti (1980) and his colleagues reviewed evidence from epidemiological studies, clinical trials, and reports of individual nurseries and concluded that newborn intensive care played a major (although unspecified) role in reducing both mortality and morbidity among low-birthweight infants. At the same time, however, the authors offered the seemingly paradoxical conclusion that, particularly among the smallest infants (those weighing 1,000 grams [2.2 pounds] or less), mortality is declining faster than morbidity; although the incidence of serious disabilities (retardation, cerebral palsy) is decreasing, the absolute number of seriously disabled persons has been increasing (Budetti et al. 1980). This finding suggests that neonatal intensive care may be contributing to the creation of future costs, including those that transcend the economic, to be borne by families, communities, and the disabled themselves.

The problem is even more acute when one considers the nature of the patient population served by the intensive-care nursery. The association between poverty and infant mortality has been well documented in the epidemiological literature. U.S. studies have also demonstrated that low socioeconomic status is associated with the birth of premature and low-birthweight infants (Rosenwaike 1971; Osofsky and Kendall 1977; Schwartz and Schwartz 1977). The greater prevalence of critically ill infants and disabled children among the poor suggests that the costs that have just been outlined will increasingly fall most heavily upon those who are least able to bear them.

The political economy and epidemiology of neonatal intensive care not only raise macrosocial issues of policy but also create a set of social, ethical, and practical dilemmas for those who reach life-and-death decisions. This is apparent when one considers yet another paradox in public policy. As our society invests its resources in neonatal intensive care, the current fiscal preoccupation with cost containment has also led to a reduction in social services—services necessary to support those disabled children who are able to survive because of neonatology's therapeutic advances. Thus, some parents confront life-and-death decisions faced with the prospects of raising disabled children in an era of diminishing social supports. The diminishing number of educational, social, and residential resources to support disabled children also places constraints on professional participants in life-and-death decisions, as the following suggests:

> Early in my research, in a case that portended the case of Baby Doe, an infant was born with Down's syndrome and duodenal atresia. A very simple surgical procedure was necessary to correct the lesion. The mother of the baby insisted that the surgery not be performed, stating that she could not care for a retarded child with a colostomy. Many members of the nursery staff felt that the surgery should be performed but were reluctant to seek a court order to perform the surgery. They feared that the mother would abandon the child to a foster home or an institution and lamented the fact that the quality of institutional care was far less than optimal.

In this case and others, those who made life-and-death decisions were brought face-to-face with larger institutional constraints.

The implications of this section for bioethical analysis are quite simple. Life-and-death decisions are not merely matters of individual conscience but, rather, take place within the context of organizations, institutions, and power relationships. The fact that decisions are a matter of politics as well as principles, interests as well as values, does not mean that bioethical analysis should abandon its traditional concerns in favor of mere pragmatism. Rather, a number of other issues (such as obedience to authority) could be incorporated into bioethical analysis. Ethical principles cannot easily be detached from the organizational, political, and economic contexts that surround them.

Conclusions

Some ethical discussions have relied upon what, from a sociological vantage point, appears to be an idealized vision of decision making in which both the medical and social parameters have been simplified. Much of ethical analysis assumes that moral choice rests exclusively with the individual, who reaches decisions apart from institutional constraints. I would like to close by suggesting, if only schematically, some of the ways in which bioethical analysis may be made more applicable to the realities faced by those who reach life-and-death decisions.

First, greater attention needs to be given to those cases in which there is little agreement as to the prognosis of the infant whose fate is being decided. There already exists an extensive literature concerning infants with congenital anomalies in which the prognoses are relatively predictable. But, with the notable exception of Jonsen and Lister (1978), all too little has been written concerning decisions under conditions of prognostic uncertainty.

Second, bioethics should attempt to adopt a logic that proceeds from actual cases to principles. The analogous logic of sociological inquiry is that of analytic induction (Znaniecki 1968; Manning 1978). To be sure, bioethical analysis already has made extensive use of the case-study method. However, cases have been selected to illustrate principles that have been formulated *a priori*. What I am suggesting here is a methodology in which principles are developed out of actual cases. Jonsen (1979) has already begun to revive and systematize the methodology of casuistry in an effort to lay the foundations for a consulting ethics. Other efforts to develop principles from cases are clearly useful to an ethics that is attuned to the realities of clinical practice.

Third, bioethical analysis should incorporate a view of the life-and-death decision as a multiparty, multimeaning transaction—an interactive model in which decisions are viewed not as individual undertakings but rather as arising out of interactions among individuals. There is a need for ethical discussion of the subtle, highly complex communication problems between staff and parents, such as those I have described.

Finally, bioethics should attempt to address the influence of the social context upon life-and-death decisions. From a sociological standpoint, the individualistic focus of bioethics lends it a reformist quality, seeking solutions in disciplined reflection rather than in changes in the social milieu. However, to the extent that the choices of decision-makers are constrained by organizational structures and ironies in public policy, changes in the broader context of decisions may be necessary for equitable decisions to ensue. If my analysis is correct, individual decisions cannot easily be extricated from larger questions of distributive justice and public policy questions of resource allocation. In addition to developing a model of the moral actor, I suggest that bioethics begin to nurture a vision of moral organizations and institutions and of a moral society.

References

Anspach, Renee R. 1987. Prognostic Conflict in Life-and-Death Decisions: The Organization as an Ecology of Knowledge. Journal of Health and Social Behavior 28, no. 3 (September):215–231.

———. n.d. Life-and-Death Decisions in Neonatal Intensive Care: A Study in the Sociology of Knowledge. Berkeley: University of California Press (forthcoming).

Becker, Howard S., Blanche Greer, Everette C. Hughes, and Anselem S. Strauss. 1961. Boys in White: Student Culture in Medical School. Chicago: University of Chicago Press.

Bogdan, Robert, Mary Alice Brown, and Susan Bannerman Foster. 1982. Be Honest but not Cruel: Staff-Parent Communication on a Neonatal Unit. Human Organization 16:10–16.

Budetti, Peter, Peggy McManus, Nancy Barrano, and Lu Ann Heinen. 1980. The Cost-Effectiveness of Neonatal Intensive Care. United States Office of Technology Assessment.

Cicourel, Aaron V. 1983. Hearing Is Not Believing. The Social Organization of Doctor-Patient Communication, Sue Fisher and Alexandra Dundas Todd, eds., pp. 221–239. Washington, D.C.: Center for Applied Linguistics.

Duff, Raymond, and R. Campbell. 1973. Ethical Dilemmas in the Special Care Nursery. New England Journal of Medicine 289:890–893.

Engelhardt, H. Tristram. 1975. Ethical Issues in Aiding the Death of Young Children. Benificient Euthanasia, Marvin Kohl, ed., pp. 180–192. Buffalo, New York: Prometheus Books.

Fox, Renee C., and Judith Swazey. 1974. The Courage to Fail. New York: Dodd-Mead.

Goss, Mary E. W. 1961. Influence and Authority Among Physicians in an Outpatient Clinic. American Sociological Review 26:39–50.

Guilleman, Jeanne Harley, and Lynda Lytle Holmstrom. 1986. Mixed Blessings: Intensive Care for Newforns. New York: Oxford University Press.

Janis, Irving L., and Leon Mann. 1977. Decision Making. New York: Free Press.

Jonsen, Albert R. 1979. The Ethicist as Consultant. Unpublished manuscript.

Jonsen, Albert R., and Michael Garland. 1976. Ethics of Newborn Intensive Care. Berkeley: Institute of Governmental Studies.

Jonsen, Albert R., and George Lister. 1978. Newborn Intensive Care: The Ethical Problems. Hastings Center Report 8:15–18.

Lo, Bernard, and Albert R. Jonsen. 1981. Bioethics and the Lack of Clinical Relevance. Unpublished manuscript.

Lorber, John. 1975. Ethical Problems in the Management of Myelomeningocele and Hydrocephalus. Journal of the Royal College of Medicine 10:47–60.

Manning, Peter K. 1978. Analytic Induction. University of Michigan. Unpublished manuscript.

McCormick, Richard. 1974. To Save or Let Die. Journal of the American Medical Association 229:171–176.

Osofsky, Howard J., and Normal Kendall. 1977. Poverty as a Criterion of Risk. Vulnerable Infants: A Psychosocial Dilemma, Jane Schwartz and Lawrence Schwartz, eds., pp. 14–29. New York: McGraw-Hill.

Parsons, Talcott. 1951. The Social System. New York: Free Press.

Ramsey, Paul. 1978. Ethics at the Edges of Life. New Haven: Yale University Press.

Robertson, John A. 1977. Involuntary Euthanasia of Defective Newborn Infants: A Legal Analysis. Vulnerable Infants: A Psychosocial Dilemma, Jane Schwartz and Lawrence Schwartz, eds., pp. 335–364. New York: McGraw-Hill.

Rosenwaike, Ira. 1971. The Influence of Socioeconomic Status on Incidence of Low Birthweight. H.S.M.H.A. Report 86:642–649.

Schwartz, Jane, and Lawrence Schwartz. 1977. Vulnerable Infants: A Psychosocial Dilemma. New York: McGraw-Hill.

Shaw, J. L. 1973. Dilemmas of Informed Consent in Children. New England Journal of Medicine 189:855–890.

Tooley, Michael. 1972. Abortion and Infanticide. Philosophy in Public Affairs 2:37–65.

Waldman, W. H. 1976. Medical Ethics and the Hopelessly Ill Child. Journal of Pediatrics 88:890–895.

Zachary, R. B. 1978. Ethical and Social Aspects of the Treatment of Spina Bifida. Lancet 2:274–276.

Znaniecki, Florian. 1968. The Method of Sociology. New York: Octagon Books.

The author acknowledges the following citations that were not incorporated into the text: "I am indebted to Prof. Albert R. Jonsen for this idea" on p. 102, line 20, following the word "*assent*"; Guilleman and Holmstrom (1986) on p. 103, line 21; Bogdan et al. (1982) on p. 104, line 32; Guilleman and Holmstrom (1986) on p. 107, line 10; Guilleman and Holmstrom (1986) on p. 112, line 25.

PART THREE

Ethical Implications of Family Formation by Surrogacy

8

The Technology of AID and Surrogacy

Kamran S. Moghissi

Abnormalities of the male reproductive system resulting in inadequate or absent semen and sperm are responsible for 30 to 40 percent of cases of infertility. Unfortunately, the management of male infertility is often ineffective, and the options for couples with this problem who do not respond to the usual therapeutic methods are limited to either adoption or artificial insemination by donor (AID). Artificial insemination, first used for humans by John Hunter, a Scottish physician, at the end of the eighteenth century, is a relatively simple procedure from a medical point of view. Donor sperm, rather than the husband's, was first used in this procedure in 1884. The development of techniques for cryopreservation of semen (the maintenance of semen by storing at very low temperatures) has expanded and simplified the use of AID.

In 1957, Schellen estimated that in the United States 100,000 babies had already been born as a result of AID and concluded, on the basis of this survey, that between 6,000 and 10,000 children are born in the United States annually as a result of AID (Currie-Cohen, Luttrell, and Shapiro 1979). As many as 300,000 U.S. children are currently believed to have been conceived by AID, and the number is growing at the rate of 10,000–20,000 births a year. In 1988, AID is offered by the majority of infertility facilities in university clinics and larger medical centers.

The principal steps in performing AID are (1) screening of the couple; (2) evaluation of the female recipient; (3) selection and management of the donor; and (4) timing and technique of insemination. A history of infertility of at least one year's duration would be sufficient grounds for initiating infertility evaluation by most physicians. Occasionally, known history of azoospermia (the absence of spermatozoa in the semen) or advanced age of the couple may be reason for accelerated infertility investigation leading to AID.

The major medical indications for AID include azoospermia or severe oligospermia (a deficiency of spermatozoa in the semen), known hereditary or genetic disorders (such as Tay-Sachs disease or Huntington's disease, which causes neurological damage), and noncorrectable ejaculatory dysfunction of the husband. Additional indications include Rh or other blood group incompatibility, spinal cord injury, exposure to environmental mutagens, presence of sperm antibodies in the male, and unexplained infertility.

Psychological Evaluation and Counseling of the Couple

The psychological aspects of AID have received very little attention. Both conscious and unconscious psychodynamic factors are present that may be beneficial or detrimental to the outcome of the procedure. Ammon and Avidan (1976) evaluated forty-four couples who received AID. Patients were interviewed by a psychologist in order to define their marital relationship, their attitude toward infertility, their feeling about AID in comparison to adoption, their attitudes toward pregnancy, the donor, and their physician. The psychological interviews of the husbands disclosed that 80 percent had guilt feelings related to their "lack of manhood." The wives also experienced guilt feelings stemming from a conflict between the feeling of sharing their husband's failure and their own pride of fertility. On the whole, however, husbands and wives unanimously favored AID over adoption, and the women wanted to experience pregnancy and delivery of a baby. The authors found the role of the psychologist in AID treatment to be important for careful selection of cases and preparation of couples to face family life in the future.

There is general agreement that the choice of AID is not necessarily reflective of neurotic needs. However, the birth of the AID child might bring forth neurotic responses ranging from mild to severe. According to Waltzer (1982), there seems to be no justification to recommend psychological or psychiatric examination routinely for couples requesting AID. But regardless of psychological evaluation, all couples requesting AID should be counseled and should sign a consent form.

Evaluation of the Female

Evaluation of the female recipient should include a thorough history, gynecological and genetic screening, physical examination, and appropriate laboratory tests, such as serologic tests for syphilis, rubella, gonorrhea smear, Papanicolaou's smear, blood type and Rh, and other tests as indicated.

Specific infertility evaluation should consist of documentation and timing of ovulation and evaluation of the fallopian tubes. Recommended techniques for ovulation timing include recording of basal body temperature (BBT), observation of cervical mucus changes, and determination of preovulatory lutenizing hormone (LH) surge by a home kit. The occurrence of ovulation may also be documented by endometrial biopsy and plasma progesterone assay. More sophisticated methods of ovulation detection and timing, such as ultrasound determination of follicular development, are costly and rarely necessary for routine clinical management.

Recording BBT

Basal body temperature recording is an inexpensive and convenient method of ovulation timing but is only of retrospective value. In practice, the time of insemination is selected to coincide with the nadir of BBT recording (immediately before the rise) of the previous cycle. Alterations of cervical mucus, however, have some predictive value.

Cervical Mucus Changes

The secretion of cervical mucus is regulated by ovarian hormones. Estrogen stimulates the production of copious amounts of watery mucus whereas progesterone inhibits the secretory activity of cervical epithelial cells. Preovulatory mucus is profuse, watery, thin, alkaline, acellular, has good spinnbarkeit (fibrosity) and ferning, and is receptive to sperm penetration. These favorable changes can readily be appreciated by trained physicians and, indeed, by patients. With coaching and proper instruction, women can perceive mid-cycle mucorrhea by the appearance of a lubricative, thin, stretchable mucus secretion at the vulva.

LH Surge

Several home kits are now available for the detection of mid-cycle LH surge that immediately precedes ovulation. In practice, the physician or the patient begin daily testing of the urine at mid-cycle. The appearance of the LH surge is identified by the development of an intense color on the dipstick.

Evaluation of Tubal Patency

Standard techniques for determination of tubal patency—whether or not the uterine tubes are open, expanded, and unobstructed—consist of the Rubin's test (uterotubal insufflation), hysterosalpingography (HSG, radiography of the uterus and fallopian tubes), and pelvic endoscopy (in which a tube with fiberoptics is inserted into the pelvic area).

In the Rubin's test, tubal patency is assessed by the passage of CO_2 through the uterus and oviducts. Because of the high rate of false positive and negative results, this test has been abandoned by most clinicians. In HSG, tubal patency is detected by the passage of a radio-opaque contrast material through the cervix, uterine cavity, and fallopian tubes. This test is more reliable than Rubin's test and also provides a permanent document relative to the appearance of the uterine cavity and tubes. It also demonstrates the presence and site of tubal obstruction. Because of these advantages, HSG is used by most clinicians as the initial screening procedure for the detection of tubal patency. Laparoscopy (in which a tube with fiberoptics is inserted through the navel in order to view pelvic organs) is reserved for those women who show an abnormality on HSG or who do not become pregnant following at least six cycles of insemination.

Selection and Management of Donors

In most centers, donors are recruited from the ranks of medical students, graduate students, or hospital residents, by word of mouth and by advertisements. The majority of physicians prefer personally to select the donors. In a survey of 711 physicians involved in AID, Currie-Cohen, Luttrell, and Shapiro (1979) found that 91.8 percent of physicians did not allow the recipients to select their own donors, and the remaining physicians did so only rarely. In the same survey, 62 percent of physicians used medical students or residents, 10 percent used other university graduate students, 18 percent used both, and the remaining 10 percent obtained donors from other sources. As a consequence, donors are not a random sampling of the general population but are a select group with presumably above-average health and intelligence.

Most centers prefer younger donors who are below the age of thirty-five and have an acceptable sperm count and quality. The criteria for acceptable sperm specimens varies among clinics. In general, sperm density of 50 million per milliliter or greater, motility above 50 percent, and normal oval forms above 60 percent are commonly used criteria. Some institutions insist that the donor be married and of proven fertility. This requirement, however, may limit the recruitment of potential donors. Other centers accept healthy young unmarried donors so long as they have high-quality semen, but make an attempt to exclude the donor if his semen does not result in any pregnancy after a predetermined number of inseminations in several recipients.

Evaluation of Donors

A thorough history is obtained from potential donors. This includes a detailed genetic history (Report of Ad Hoc Committee 1982). Those with a family history of mental retardation, neurologic disorders, unexplained death under thirty, and significant congenital defects are excluded. Recognized and suspected hereditary and familial disorders such as diabetes and hypertension should be noted and avoided when matching the recipient (Report of Ad Hoc Committee 1982). In rural communities, the residence of the donor's family should not be in close proximity to that of recipient. A physical examination is performed on the donor and those with detected or suspected congenital abnormalities are excluded. At the time of examination, careful note is also made regarding the donor's height, weight, color of hair and eyes, complexion, body build, racial and religious background, and blood type.

In addition to semen analyses, the American Fertility Society (AFS) Ad Hoc Committee on Artificial Insemination (Report of Ad Hoc Committee 1982) has recommended the following laboratory examinations: Hepatitis B antigen (HBSag); serologic test for cytomegalovirus and syphilis; cultural studies for *Neisseria gonorrhea* and *Chlamydia trachomatis;* and blood type and Rh factor. Tay-Sachs screening test for all Jewish donors and electrophoretic evaluation for sickle-cell trait for black donors is recommended. A karyotype is not recommended unless there is a specific indication. In a few centers, chromosome analysis is performed on all prospective donors (David and Lansac 1979; Rioux and Ackman 1979). When fresh semen is used, an initial semen screening for HTLV-III virus antibodies (indicating acquired immune deficiency syndrome—AIDS) should be obtained. A positive assay should be verified with a Western Blot test (the most specific test for AIDS virus) prior to notifying the potential donor. When frozen semen is to be used, the donor should be screened initially for HTLV-III, and if the test is negative, semen samples may be collected for cryopreservation. The donor should be tested again in sixty days for HTLV-III and the specimen released for use only if the results are negative. It is recommended further that all tests for sexually transmitted diseases be repeated at least every six months.

Limitations in the Use of Donors

In some AID programs, a few donors are used for long periods and often for large numbers of women. In certain cases, this may create a substantial risk of consanguinity. The concern is the possibility that AID children conceived from the same donor—who are, therefore, half-brother and half-sister—might unwittingly marry and have children. The progeny of first-cousin marriages show a higher than normal mortality and congenital malformation rate; half-brothers and half-sisters are more closely related genetically than first cousins, so their children would be at a still greater disadvantage (McLaren 1973).

The possibility of consanguineous marriages, however, is very small provided that the number of babies fathered by a single donor remains small. Glass (1960) has estimated that if 2,000 live children a year were to be born in Great Britain as a result of successful use of AID and if each donor was responsible for five children, an unwitting incestuous marriage is unlikely to take place more than once in 50 to 100 years.

If the number of children from each donor is strictly limited, not only are the risks of incestuous marriage minimized, but also any demographically undesirable effects of donor selection are avoided. A further advantage is that there will be less risk of disproportionate spread of any harmful recessive genes that the donor may be carrying (McLaren 1973). In France, the Center for the Study and Preservation of Semen (CECOS, Centre d'Etude et de Conservation du Sperme), a network of university hospital centers, obtains only a limited number of ejaculates, usually five or six from a given donor during a brief period of about one month (David and Lansac 1979).

In practice, by avoiding the matching of a donor with recipients from the same area of family origin, the risk of consanguinity between offspring of recipients by the same donor can be virtually eliminated. In general, a donor should not be used after fifteen successful pregnancies.

Compensation to Donors

Some centers, such as CECOS in France, are able to solicit semen donations without payment to donors (similar to blood donation). In the United States, Currie-Cohen, Luttrell, and Shapiro (1979), in their survey of AID centers, found that virtually all respondents paid donors for their services. The payment varied from one institution to another. Almost half paid $25 per ejaculate, and 88 percent paid between $20 and $35. The AFS recommends that any written agreement between a sperm bank or physician and donor should indicate that any payment to the donor is to provide for transportation, inconvenience, and other incidental expenses incurred in donating the sample, rather than direct payment for semen.

Matching

In most AID programs, attempts are made to match the physical appearance of the donor with that of the recipient's husband. The criteria for matching include: race, body build, height, complexion, color of hair and eyes, and possibly blood type and Rh. Some centers go as far as matching for such factors as ethnic and religious background and educational level. It must be realized, however, that with an increased number of characteristics to match, greater difficulties will be encountered in finding suitable donors for any given couple.

Technique of Insemination

Insemination should be timed as closely as possible prior to the time of ovulation as determined by BBT, cervical mucus changes, and other indicated tests. Because ovulation cannot be timed accurately, most physicians prefer to perform several inseminations per cycle. It is believed that sperm can survive for at least forty-eight hours in the female reproductive tract and maintain its fertilizing ability. Thus, the majority of centers perform multiple inseminations, usually two per cycle at two-day intervals (Behrman 1979; Behrman 1968).

To perform artificial insemination, the semen is obtained from the donor and is allowed to liquefy. The woman recipient is placed in the lithotomy position. A speculum is inserted in the vagina, and the cervix is wiped clean with a dry cotton ball. Two techniques of insemination are most commonly practiced (Behrman 1979; Behrman 1968):

1. Intracervical: Approximately 0.1 milliliters of the semen is injected with an inseminating polyethylene tube into the cervical mucus. The rest spills out and collects in the vaginal pool. Some physicians insert a special insemination plug in the vagina to prevent the discharge of the semen from the vagina, but others do not find this to be necessary. The patient is instructed to rest on the examining table for thirty minutes before leaving the room.

2. Cervical cap: 0.1 milliliters of the semen is injected in the cervical canal, and the rest is placed in a cervical cap that the woman removes later on.

Semen Collection and Preservation

Semen should be collected by masturbation, after the donor has abstained from sexual activity for three to five days. The collection should occur at a convenient location in the vicinity of the sperm bank and be processed immediately. Most sperm banks add glycerol (7–10 percent V/V) as cryopreservative to the semen (Foote 1982; Sherman 1978). The sample is then placed in plastic straws or small ampules and suspended in liquid nitrogen vapor to freeze. Some centers use an automated and controlled freezing chamber, but data are insufficient to document the superiority of this technique. The samples are then preserved in liquid nitrogen at a temperature of −196°C (−476°F). Prolonged preservation seems to be deleterious to sperm survival. When required, straws or ampules are placed in a water bath at room temperature for thawing within five minutes (Foote 1982; Sherman 1978).

Cryopreservation of semen was first attempted in animal husbandry and has led to great improvement in efficiency of livestock production. An important factor responsible for achieving these results was the development of techniques to harvest and preserve semen and inseminate females under conditions that would assure optimal conception rates. With the discovery of the cryoprotectant properties of glycerol, modern cryobiology was born.

The first AID pregnancy resulting after use of frozen semen was reported by Bunge, Keettel, and Sherman (1954). In 1978, Sherman (1978) surveyed the literature and reported that the use of frozen-stored human semen had resulted in 1,464 normal children, 11 abnormal children, and 113 spontaneous abortions. The incidence of abnormal children and spontaneous abortions among the cases was lower than in the naturally reproducing population.

Since this report, a large number of sperm banks have been established in the United States and other countries, resulting in more pregnancies from the use of frozen semen. The major advantages of frozen semen are the immediate availability and the convenience in timing insemination. Also, there is greater possibility of matching donors. These advantages are partly negated by the lower fertility rate and increased cost (Steinberger and Smith 1973).

The selection of donors for sperm banking is similar to that for fresh semen. However, the semen of each donor should be tested for its freezing ability. It is well documented that semen obtained from certain donors does not withstand the freezing process despite initial adequate sperm density and motility. Extensive loss of motility of spermatozoa in a sample as a result of freezing is associated with a considerable decline in pregnancy rates subsequent to insemination.

Evaluating AID Results

The success of AID depends on three main factors: first, the recipient woman's fertility potential; second, the quality and potential fertilizing ability of semen used; and third, the timing and frequency of insemination. Female fertility potential is reduced by disorders of ovulation, organic diseases of the reproductive tract, and increasing age of the recipient.

Pregnancy Rate

In discussing the pregnancy rate resulting from a given program, one needs to consider the size and type of population treated, the thoroughness with which these factors have been evaluated, and the experience and expertise of the center at which the patients are treated. The only valid way of assessing the success of an AID program is to estimate the pregnancy rate. Elimination of patients over age forty and those with additional infertility factors make the success rates about 70–75 percent when fresh semen is used (Berquist et al. 1982; Glezerman 1981; Smith, Rodriguez-Rigau, and Steinberger 1981; Virro and Shewchuck 1984). In various studies, rates as high as 90 percent and as low as 60 percent have been reported for patients completing at least six cycles of treatment (Steinberger and Smith 1973). With frozen semen, the success rate is about 10–20 percent lower and ranges from 40 to 70 percent (Empraire 1982; Friedman 1977; Mahadevan et al. 1982; Schoysman and Schoysman-Deboeck 1976; Steinberger and Smith 1973; Trounson et al. 1981).

It has been stressed that the overall success rate is meaningless unless the number of treatment cycles is taken into account. Higher treatment cycles are usually associated with a higher rate of success. For example, Virro and Shewchuck (1984), using fresh semen, found that in those patients who completed three insemination cycles, a pregnancy rate of 40 percent was obtained; a pregnancy rate of 92 percent was recorded when six insemination cycles were completed. Similarly, Friedman (1977) reported an overall pregnancy rate of 40 percent for patients inseminated with frozen semen. In the same study, when women received at least four cycles of insemination, the pregnancy rate rose to 56 percent.

The reported spontaneous abortions in pregnancies achieved by fresh semen varies between 4.1 and 17.8 percent with 13.8 percent as a mean. The respective figures for frozen semen are 13.0–26.6 percent with the mean 16.1 percent (Alfredsson, Gudmundsson, and Snaedal 1983). The abortion rate tends to be higher with advancing age, with the use of frozen semen, and when clomiphene citrate is used for ovulation induction. A few reports show the male-to-female sex ratio at birth to be altered, but this result is not borne out by studies of large groups of AID recipients, which indicate no significant change.

Complications

Generally, AID is a safe procedure. However, two important complications need to be considered. First, gonorrhea and other sexually transmitted diseases (STDs) may be transmitted by semen obtained from infected symptomatic and asymptomatic donors (Fiumara 1974; Jennings, Dixon, and Nettles 1977). Because of this complication, some investigators (Sherman 1978) recommend using only cryopreserved ejaculates that, after appropriate cultures, have been found to be free of *N. gonorrhea.* Until recently, it was believed that *N. gonorrhea* was rapidly destroyed by air exposure, but several investigators have demonstrated that *N. gonorrhea* in contaminated semen does not lose viability in the time frame during which most specimens are used (Jennings, Dixon and Nettles 1977). In fact, *N. gonorrhea* may even survive cryopreservation. Routine culture of semen will add to the cost of the procedure but has been practiced by some institutions. Transmission of other STDs by AID has not been documented but is a distinct possibility. In all cases, the couple should be advised of this possible complication.

Second, transmission of genetic and hereditary disorders, though extremely uncommon, also need to be considered. Most centers involved in AID screen their donors for genetic disorders; only a few perform routine karyotyping (David and Lansac 1979). Carrier testing of the donor is appropriate if the donor or recipient belongs to a defined population with a higher frequency of a specific trait (such as blacks, for sickle-cell trait). Unfortunately, extensive screening of the donor will inevitably drive up the cost of the procedure and must be tempered against the yield of a positive finding.

Follow-up of Children Born as a Result of AID

Because donors for AID are usually preselected and screened for their physical and mental fitness, AID children may be expected to be physically and mentally superior to average children. Iizuka et al. (1968) evaluated the physical and mental development of fifty-four children born as a result of AID. The distribution of IQs of those children who were two and a half years of age and older was found to be in a higher range than that of controls. The physical and mental development of children following heterologous insemination in no case was inferior to that of the control group resulting from natural impregnation. Freeze preservation of the semen did not affect the IQ of the offspring.

Factors Influencing the Success of AID

The available data (Behrman 1968; Schoysman and Schoysman-Deboeck 1976) indicate that a monthly schedule of two inseminations is superior to a single insemination. Increasing the number of inseminations per cycle beyond two does not seem to increase the success rate and adds to the cost.

A frequent question asked by the couple is whether one can mix the husband's sperm with the donor sperm. An answer to this question is provided by the studies of Quinlivan and Sullivan (1977), who showed almost a doubling of pregnancy rate when he asked the couple to avoid coitus for three days before insemination. Also the pregnancy rate following AID has been found to be significantly higher in patients whose husbands were azoospermic than in women whose husbands were severely subfertile (Empraire, Gauzere-Soumireu, and Audebert 1982).

Younger donors seem to have a better chance of impregnating the recipient. Donors less than twenty years of age initiated a higher pregnancy rate when compared to older donors (Mahadevan et al. 1982).

Peak fertility for females occurs between twenty-one and twenty-four years of age and will gradually decline thereafter. Several reports indicate that patients over age thirty to thirty-five who are receiving AID treatment have lower pregnancy rates, require a greater number of inseminations to achieve pregnancy (Albrecht, Cramer, and Schiff 1982; Virro and Shewchuck 1984), and have a higher spontaneous abortion rate (Virro and Shewchuck 1984).

The presence of other factors interfering with the normal reproductive process in the female will invariably reduce the pregnancy rate. Smith, Rodriguez-Rigau, and Steinberger (1981) found that ovulatory dysfunction, even after treatment, increased the number of insemination cycles required before conception occurred. Similarly, the presence of tubal disease, endometriosis, and cervical mucus abnormalities has been reported to be associated with lower fertility.

Some authors have suggested the use of clomiphene citrate to bring about predictable and regulated ovulation and to allow time to schedule the insemination when office personnel and donors are readily available (Klay 1976). This technique may offer an advantage to patients who have few ovulations or have irregular ovulation. However, as clomiphene citrate may have adverse effects on cervical mucus, it is not recommended for routine use.

Legal Aspects of AID

Statutes

Couples considering AID are usually unaware of the legality of this procedure and need to be advised regarding the child's legal status. There is, at present, little regulation in many states concerning the legality of the procedure and the status of the child (Shaman 1980; Virro and Shewchuck 1984). However, many states have begun to address this issue. Six states (California, Oklahoma, Virginia, Washington, Alaska, and Oregon) have statutes that limit the use of AID to practicing physicians. A child conceived through AID is still in legal limbo in twenty-two states. The other twenty-eight states have enacted laws specifying that a child conceived by means of artificial insemination carried out with the consent of the husband is the legal offspring of the couple (Alabama, Alaska, Arkansas, California, Colorado, Connecticut, Florida, Georgia, Idaho, Illinois, Kansas, Louisiana, Maryland, Michigan, Minnesota, Montana, Nevada, New Jersey, New York, North Carolina, Oklahoma, Oregon, Tennessee, Texas, Virginia, Washington, Wisconsin, and Wyoming). Fifteen of these states also specifically provide that a man who furnishes sperm for the artificial insemination of someone other than his wife is not the child's legal father (Alabama, California, Colorado, Connecticut, Idaho, Illinois, Minnesota, Montana, New Jersey, Nevada, Oregon, Texas, Washington, Wisconsin, and Wyoming) (Donovan 1986). Five states (Georgia, Kansas, Oklahoma, Oregon, and Washington) require that the consent be filed with a state agency.

When litigation has occurred, the courts have usually held that the AID child is legitimate and entitled to support from his mother's husband at the time the procedure took place. In most jurisdictions, AID is not considered adultery unless there has been actual sexual intercourse. The courts have never held an anonymous sperm donor responsible for supporting the child. In California, for example, if a wife is inseminated artificially with semen of a man who is not her husband, under the supervision of a licensed physician and with the consent of her husband, the husband is treated in law as if he were the natural father of the child. Unfortunately, most jurisdictions have not simplified the issue so well, and it behooves physicians or institutions who plan to perform AID to determine its legal status in their state or city.

Only the states of Oregon and Idaho and New York City require that donors be screened for genetic abnormality or infectious disease, and even these laws simply rely on information supplied by the donor to determine whether he has a risky medical history (Donovan 1986).

Informed Consent

The issues directly applicable to physicians are, principally, those related to the performance of AID. To avoid the most obvious entanglement, informed consents from all parties are mandatory. Consent forms should cover at least the following: (1) that more than one insemination may be required; (2) that despite multiple inseminations, conception cannot be guaranteed; (3) that reasonable means will be used to select donors; (4) that the patient and her husband forever relinquish the right to discover the identity of the donor.

Record Keeping and Confidentiality

Thus, in most institutions, AID is highly secretive, both because of ambiguities in its legal status and to protect donors. Unless the anonymity of the donor is adequately maintained against future legal, financial, and emotional claims, donor recruitment will become extremely difficult, if not impossible. The issue of confidentiality of the donor is a very important one. Anonymity protects the donor from legal involvement in the legitimacy and inheritance rights of children born through AID. The recipient should also be protected against legal claims to her child by the donor.

Many concerns for donor anonymity could be resolved by legislation clearly defining AID and the role of the participants. Until such legislation is enacted, the majority of physicians feel the donor must remain anonymous and the social father's name should be recorded on the birth certificate, as society views him as the real father. Confidentiality is also reflected in the incomplete records kept by many physicians. Such secrecy can obviously impede a critical examination of AID, its consequences, and possible complications. Finally, the issue of whether or not to inform the child conceived via AID of his/her origin, as is done with adoptees, is unresolved. The predominant view is that a child conceived by AID should not be informed (Ammon and Avidan 1976; Behrman 1968).

Other Issues

AID and Single Women

Many doctors refuse to perform AID upon single women, and some nineteen states prohibit it by statute. When doctors act independently they are not engaged in state action and, hence, are not subject to relevant constitutional strictures. However, the statutes are determined by the state and are, therefore, subject to constitutional question. Judging by prevailing case law, there is some doubt that statutes prohibiting single women from obtaining AID are constitutional (Shaman 1980). The single woman has obviously the constitutional right to bear a child. However, in the absence of specific statutes, many issues—such as the status of the child, the right of the mother to bring a civil action for child support against the physician and/or the institution where the procedure was performed, and the right of the child to estate inheritance from the donor—remain unresolved.

Strong and Shinfeld (1984) reviewed the relevant social science literature on AID in single women. Their findings suggested that the decision to perform the procedure should be made on a case-by-case basis. The approach they recommended was to use the following factors as guidelines for dealing with a request for AID by a single woman: (1) health of the prospective recipient; (2) the recipient's financial situation and whether she is capable of adequately supporting a child; (3) counseling of the potential recipient concerning the importance of child-adult interaction, in which the counselor shows concern about the need for such interaction and a commitment to seeing that it is provided; (4) the presence of a support system, such as immediate or extended family. In those cases in which AID is performed with fresh semen, the donor should be informed that the intended recipient is a single woman. With these provisions, the authors felt that AID for single women is permissible but that the physician has a right to refuse to carry out such requests.

AID in Surrogate Women

This method involves artificial insemination of a fertile woman with the sperm of the husband of an infertile woman. The surrogate, who has been employed for this purpose, agrees to bear a child for the infertile couple and turn it over to them at birth, either by giving the child up for adoption or by relinquishing her parental rights and obligations.

This is a much more socially problematic practice than AID in either married couples or single women because it raises new issues of maternity involving the identity of the rearing mother and the commercialization of motherhood. The maternity issue involves the surrogate's ability and right to change her mind and keep the child. It also includes the possibility that the sperm donor may reject a defective child. Because of such unresolved legal and ethical issues many physicians are reluctant to become involved in insemination of surrogate mothers (Elias and Annas 1986).

Conclusion

In the absence of effective treatments for many conditions leading to male infertility, AID provides an alternative method of achieving parenthood for many couples who may otherwise be deprived of the pleasure of rearing a child. The procedure is safe, simple, highly successful, and acceptable to many couples. Legal issues regarding many aspects of AID and the child's status remain unresolved in many states. The procedure has been sanctioned by most ethical experts but prohibited by some religious groups. Concerns regarding the possibility that the procedure might create psychological problems in the husband, wife, and/or donor exist and need further investigation. A similar concern is that psychologic damage may result if the child learns inadvertently that AID was responsible for his or her origin.

Unfortunately, long-term, well-designed prospective studies are neither available nor easy to establish. The consensus of available reports and medical experts is that the medical, social, and psychological benefits of AID considerably outweigh any hypothetical risks. In fact, available information indicates that for the overwhelming majority of couples, the outcome of AID is a happy and rewarding experience.

References

Albrecht, Bruce H., Daniel Cramer, and Isaac Schiff. 1982. Factors Influencing the Success of Artificial Insemination. Fertility Sterility 37:792–797.

Alfredsson, Jon H., S. P. Gudmundsson, and G. Snaedal. 1983. Artificial Insemination by Donor with Frozen Semen. Obstetrics and Gynecology Survey 38:305–313.

American Fertility Society. 1982. Report of the Ad Hoc Committee on Artificial Insemination. Birmingham, Alabama.

Ammon, D., and D. Avidan. 1976. Artificial Insemination Donor: Clinical and Psychologic Aspects. Fertility Sterility 27:528–532.

Behrman, Samuel J. 1968. Technique of Artificial Insemination. Progress in Infertility, S. J. Behrman and R. W. Kistner, eds., pp. 98–111. Boston: Little Brown.

———. 1979. Artificial Insemination. Clinical Obstetrics and Gynecology 22:245–253.

Berquist, S. J., J. A. Rock, J. Miller, D. S. Guzick, A. C. Wentz, and G. S. Jones. 1982. Artificial Insemination with Fresh Donor Semen Using the Cap Technique: A Review of 278 Cases. Obstetrics and Gynecology 60:195–199.

Bunge, R. G., W. C. Keettel, and J. K. Sherman. 1954. Clinical Use of Frozen Semen. Fertility Sterility 5:520–529.

Currie-Cohen, Martin, Lesleigh Luttrell, and Sander Shapiro. 1979. Current Practice of Artificial Insemination by Donor in the United States. New England Journal of Medicine 300:585–590.

David, G., and J. Lansac. 1979. The Organization of the Centers for the Study and Preservation of Semen in France. Human Artificial Insemination and Semen Preservation, G. David and W. S. Price, eds., pp. 15–26. New York: Plenum Press.

Donovan, Patricia. 1986. New Reproductive Technologies: Some Legal Dilemmas. Family Planning Perspectives 18:57–60.

Elias, Sherman, and George J. Annas. 1986. Social Policy Consideration in Noncoital Reproduction. Journal of the American Medical Association 255:62–68.

Empraire, Jean-Claude, Elisabeth Gauzere-Soumireu, and Alain J. M. Audebert. 1982. Female Fertility and Donor Insemination. Fertility Sterility 37:90–93.

Fiumara, N. J. 1974. Transmission of Gonorrhea by Artificial Insemination. British Journal of Venereal Disease 48 (1972):308–309.

Foote, Robert H. 1982. Cryopreservation of Spermatozoa and Artificial Insemination: Past, Present and Future. Journal of Andrology 3:85–100.

Friedman, Stanley. 1977. Artificial Donor Insemination with Frozen Human Semen. Fertility Sterility 28:1230–1233.

Glass, D. V. 1960. Report of the Departmental Committee on Human Artificial Insemination. London: H. M. Stationery Office.

Glezerman, Marek. 1981. Two Hundred and Seventy Cases of Artificial Donor Insemination Management and Results. Fertility Sterility 35:180–187.

Iizuka, Rihachi, Yoshiaki Sawada, Nobuhiro Nishina, and Michie Ohi. 1968. The Physical and Mental Development of Children Born Following Artificial Insemination. International Journal of Fertility 13:24–32.

Jennings, Richard T., Richard E. Dixon, and John B. Nettles. 1977. The Risk and Prevention of Neisseria Gonorrhea Transfer in Fresh Ejaculate Donor Insemination. Fertility Sterility 28:554–556.

Klay, Leonard J. 1976. Clomiphene-Regulated Ovulation for Donor Artificial Insemination. Fertility Sterility 27:383–388.

Mahadevan, Maha M. et al. 1982. Influence of Semen and Donor Factors on the Success Rate of Artificial Insemination with Frozen Semen. Clinical Reproductive Fertility 1:185–193.

McLaren, A. 1973. Biological Aspects of AID. Law and Ethics of AID and Embryo Transfer, 1st edition, pp. 3–9. New York: Elsevier, Excerpta Medica.

Quinlivan, W.L.G., and H. Sullivan. 1977. The Immunologic Effects of Husband's Semen on Donor Spermatozoa During Mixed Insemination. Fertility Sterility 28:448–450.

Rioux, Jacques E., and C.D.F. Ackman. 1979. Artificial Insemination and Sperm Banks: The Canadian Experience. Human Artificial Insemination and Semen Preservation, G. David and W. S. Price, eds., pp. 31–34. New York: Plenum Press.

Schellen, A.M.C.M. 1957. Artificial Insemination in Humans. Amsterdam: Elsevier.

Schoysman, R., and A. Schoysman-Deboeck. 1976. Results of Donor Insemination with Frozen Semen: Sperm Action. Progress in Reproductive Biology 1 (1976):252.

Shaman, Jeffrey M. 1980. Legal Aspects of Artificial Insemination. Journal of Family Law 18:331–349.

Sherman, J. K. 1978. History of Artificial Insemination and the Development of Frozen Semen Banks. The Integrity of Frozen Spermatozoa, A. P. Ruifret and J. C. Petricciani, eds., pp. 201–211. Washington, D.C.: National Academy of Sciences.

Sherman, J. K., and J. Rosenfield. 1975. Importance of Frozen-Stored Human Semen in the Spread of Gonorrhea. Fertility Sterility 26:1043–1047.

Smith, Keith D., Luis J. Rodriguez-Rigau, and Emil Steinberger. 1981. The Influence of Ovulatory Dysfunction and Timing of Insemination on the Success of Artificial Insemination Donor (AID) with Fresh and Cryopreserved Semen. Fertility Sterility 36:496-502.

Steinberger, Emil, and Keith D. Smith. 1973. Artificial Insemination with Fresh or Frozen Semen. Journal of the American Medical Association 223:778–783.

Strong, Carson, and J. S. Schinfeld. 1984. The Single Woman and Artificial Insemination by Donor. Journal of Reproductive Medicine 29:293–299.

Trounson, Alan O., C. D. Mathews, G. T. Kovacs, A. Spiers, D. M. Steigrad, D. M. Saunders, W. R. Jones, and S. Fuller. 1981. Artificial Insemination by Frozen Donor Semen: Results of Multicenter Australian Experience. International Journal of Andrology 4:227–234.

Virro, M. R., and A. B. Shewchuck. 1984. Pregnancy Outcome in 242 Conceptions After Artificial Insemination with Donor Sperm and Effects of Maternal Age on the Prognosis for Successful Pregnancy. American Journal of Obstetrics and Gynecology 148:518–524.

Waltzer, Herbert. 1982. Psychological and Legal Aspects of Artificial Insemination (AID): An Overview. American Journal of Psychotherapy 36:91–102.

9

Secrecy and the New Reproductive Technologies

Judith N. Lasker and Susan Borg

Secrecy is often a central issue in discussions of artificial insemination with a donor's sperm (AID), the oldest of the alternative methods of conception and one which has been used by a few hundred thousand families worldwide (Curie-Cohen, Luttrell, and Shapiro 1979; Annas 1979). The persistence of secrecy with the use of AID presents a number of difficult ethical and psychological dilemmas for the families. The purpose of this paper is to compare attitudes toward secrecy among those who use AID and people who are trying more recent methods such as in vitro fertilization, surrogate motherhood, and ovum transfer. We will consider why secrecy has continued to be a key feature of AID and not of any other method of conception. We suggest that a key characteristic which differentiates AID from other methods—its use exclusively for male infertility—provides most of the explanation. This is because male infertility is a condition with greater stigma attached to it than female infertility. The reasons for this difference will be explored.

The data for this paper come from several sources: thirty semi-structured interviews carried out in person and by phone with men and women who are trying or have tried one of the alternative methods of conception, twenty-seven interviews with professionals who offer those methods, and a mailed questionnaire survey of 85 men and women who are considering or have tried one of the methods. The interview sample of people using

This paper was presented at the Society for the Study of Social Problems, August 1987. It is based on research carried out for *In Search of Parenthood: Coping with Infertility and High-Tech Conception* (Beacon Press, 1987) and uses portions of that book with permission of the publisher. The research was made possible by several grants from Lehigh University with the assistance of Linda Wimpfheimer and Mary Ann Hughes. Thanks to Louise Potvin, Kandi Stinson, and Brenda Seals for their comments on an earlier version of the paper.

the methods was obtained partly by word of mouth and referral from providers; the rest of the interviewees and all respondents to the mailed questionnaire were people who answered a request we placed in the national and local newsletters of RESOLVE, an organization for infertile people. This is not a representative group, but rather it is skewed toward people who have demonstrated a willingness to talk about their infertility and to seek support from others. Nonetheless, as we discovered, secrecy was still an issue for many of these respondents.

Secrecy with AID

Secrecy is a key issue for people involved in donor insemination. It is almost always surrounded by secrecy—secrecy about the identity of the donor, secrecy between the couple and their friends and family about what they are doing, and secrecy from the child conceived about his or her true origins. The first successful AID on record was carried out in 1884; it was done during an examination of the woman without even telling the couple involved. Since then, as British sociologist Robert Snowden points out, some physicians have advocated, if possible, not telling the husband of his infertility or that a donor is going to be used (Snowden, Mitchell, and Snowden 1984).

Most physicians and other staff members of AID programs strongly encourage their patients to keep AID a secret from everyone (Beck and Wallach 1981). Some even suggest not telling the obstetrician who delivers the baby, so that the husband's name will be put on the birth certificate without hesitation. In addition, most studies which ask couples their intentions regarding secrecy find that the great majority intend to tell no one, and very few expect to tell the child (Leeton and Blackwell 1982; Ledward, Crawford, and Symonds 1979; Czyba and Chevret 1979; Milsom and Bergman 1982; Rowland 1985). Thus professionals and AID users concur that the procedure is best forgotten. In our survey of unusually open people—RESOLVE members who had volunteered to fill out a questionnaire—62% felt that it was important to keep their use of AID secret, and half either would not tell the child or were uncertain about whether they would.

The advice of one psychiatrist on the issue demonstrates how strongly secrecy has been emphasized in the practice of AID:

> For the child's sake, I prefer that absolutely nobody but the parents themselves know of the insemination therapy. I feel strongly that under no circumstances should they, or need they, ever tell the child the method of conception—in fact they should forget about it themselves (Waltzer 1982).

No matter how strongly they express a commitment to secrecy, most of the people in our study and at least a substantial minority in other studies had told someone, if only one or two others. They are selective, often telling one set of parents, but not the other, or certain friends they assume will be sympathetic. But they have found it very difficult not to tell anyone at all (Manuel, Chevret, and Czyba 1980).

Secrecy with Other Methods

For people trying in vitro fertilization (IVF), secrecy is not much of an issue. Almost everyone we surveyed who has or is trying to have a child through IVF intends to tell the child. In fact, they take exactly the opposite approach from people using AID. They generally *want* the children to know about their IVF origins so they will realize that their parents went through a great deal in order to have them. Some parents recognize that they must proceed with caution, as this mother of an IVF child points out:

> I want her to know what we went through so she will understand how special she is to us. I think you can get into a bind—you don't want it to be something she has to live up to.

Couples who hired surrogates were a little less sure, but they also generally planned to treat the subject openly. Surrogate birth is similar to adoption because of the existence of a birth mother who relinquishes the child to a couple through a legal adoptive process. Not surprisingly, most parents expect to tell their children about their legal mother in the same way professionals now advise parents to talk about adoption—gradually, openly, and early in the child's life.

The contrast between AID and other methods is very striking. We are not aware of any other study which asks people about their attitudes toward sharing the experience of in vitro fertilization or surrogate motherhood with friends and with the child. In our analysis of the questionnaire data, we found that those who had used or were in the process of using AID were significantly more likely than the others to want to keep this friends and from the child; people who had tried IVF, on the other hand, were much more likely than all others to want to tell other people and to tell the child. Artificial insemination with husband's sperm (AIH) falls in the middle (see Table 9.1). The survey sample included no one who had tried surrogate motherhood; in our interviews, however, there were eight people who had. Although one couple expressed caution about mentioning it to people in their small conservative rural community, no one in this group considered secrecy from family members or from the child desirable. We

TABLE 9.1

Attitudes Toward Secrecy by Method Tried

	Important to keep secret from others			Expect to tell children		
	Yes	Uncertain	No	Yes	Uncertain	No
AID	62.1%	3.4%	34.5%*	50.0%	25.0%	25.0%**
IVF	22.6	6.5	7.0*	96.4	0	3.6**
AIH	36.0	12.0	52.0	68.2	22.7	9.1

*p<.05

**p<.01 (Comparison of users of each method to all nonusers of that method in the sample)

did learn from surrogate program staff of a few cases where women had pretended to be pregnant at the same time as the surrogate mother, but these were described as exceptions.

The Parents' Reasons for Secrecy with AID

Why the tremendous difference in attitude? One reason offered by people involved with AID is that it is much easier to keep it secret than other forms of conception, and certainly easier than adoption. Unlike the situation of adoption, where others know and neither of the parents is involved genetically, the child born of AID appears to be the product of the parents' normal pregnancy and birth. There is no legal transaction, the father's name is on the birth certificate, and often the attending obstetrician or midwife is unaware of AID. Certainly most parents who conceive "naturally" do not tell children about the circumstances of their conception.

Another reason given for secrecy is the fear that the child will be stigmatized by others and ridiculed by his or her playmates. The parents are also afraid that other family members will disapprove of AID and therefore reject the child. Of the parents who used IVF, 6.3% wrote that they wanted to keep it secret in order to avoid public disapproval; 30% of the AID parents expressed this attitude. This is a very real concern, particularly in light of greater public opposition to AID than to IVF (Lasker and Borg 1987). As one woman told us,

> The doctor told us it's nobody else's business, that it's just between the two of us and him, and no one else has to know. That was a hard thing, you know, because so many times you want to say it. It's a hard thing to

> explain, and people seem to think that if you can't have your own, you shouldn't have any or you should adopt. I have a feeling his parents wouldn't accept the baby if they knew. They wouldn't think she was a part of him.

Although the problem of public or family approbation may be faced as well by parents who adopt or have children from surrogate mothers, with AID one might argue that the method allows that exposure to public disapproval to be avoided, while adoption and surrogate motherhood are much more difficult to conceal.

Some parents say they do not want to tell the child because they know it will be impossible to locate the genetic father. Therefore, they would prefer to protect the child from the frustration of having no information about half of his or her genetic heritage.

Arguments Against Secrecy with AID

The parents' explanations center on protection of the child, an understandable motivation. If the child is not to find out, they reason, then no one must be told. Yet some observers argue that the child may actually be harmed by such secrecy (Berger 1980). Some claim that it is essential that children know their medical history and not assume they may share any medical problems experienced by the fathers who raise them (Annas 1979). Others also fear that maintaining a secret about the child's origin is ultimately unhealthy for the *family* (Manuel, Chevret, and Czyba 1980). Dr. Aphrodite Clamar, writing in the *American Journal of Psychoanalysis* (1980), emphasizes this possibility: "By its very nature, a secret is a potent force, assuming undue proportion and power within the family—an existential fact that remains unspoken, yet controls and colors the lives of the people involved."

Many of the people in our study felt very torn between a desire to keep the information private and a strong need to talk to others about it. They are fearful of others' reactions. Yet they also talked about the stress of "living a lie" and their wish to share such an important event with those to whom they are close. Some wish they could meet other people who also used AID to compare their feelings and experiences with them.

The secrecy inevitably produces some awkwardness at times, especially when family members and friends speculate, as they always do, about which parent a child resembles. One father recalled:

> When he was little, everyone said he looked just like me. My mother has a picture of me when I was little and she held it up against him and she said they looked identical, like twins. I used to say, "No, look how different he looks," but now I just play along with it—I know what really happened.

Writers such as Clamar worry that children may sense that something is being hidden from them. Those who are eventually told often feel relieved because they had guessed that there was something different, or bad, about themselves (Snowden, Mitchell, and Snowden 1984). The potential for divulging the secret is always present, since many couples do tell a few friends or family members, and the mother's medical records may include information about AID. If the children do find out from others, they will learn that their parents can deceive them about something so central to their identities. If they do not find out, they will assume a medical history which is false and risk marrying incestuously.

When parents have told their children, it has often been due to special circumstances, such as a later divorce and custody battle or when the children faced infertility themselves as adults. An important follow-up study by Snowden, Mitchell, and Snowden (1984) of English families who had used AID found only a few who had told of their offspring of the AID origins, and only as adults. In each case, the mother had wanted her son to know he was not genetically related to his father. The husbands in these families were disabled, immoral, or economic failures, and the mother's revealing of the AID can be seen as an act of hostility. In fact it is likely that AID information, kept secret for so many years, may be revealed in an angry environment. Dissatisfied fathers may blurt out to an unruly child, "You're not mine anyway." Studies of adoptees have also found that many were told in an angry way or at an inappropriate time, with damaging results (Lifton 1979).

On the other hand, Snowden and colleagues also interviewed adults who had been told of their AID origins. They all said that they had suspected something all along, and that the telling had been a relief. They also said they felt especially important, and that their relationship to their father was enhanced by realizing what he had been through (Snowden, Mitchell, and Snowden 1984).

Another Explanation for Secrecy with AID: Protecting the Fathers

The explanation for secrecy which is not often offered explicitly is the desire to protect the two fathers, both the sperm donor and the social/legal father. Some physicians indicate a preference for secrecy based on a concern that it will be impossible to recruit sperm donors without a guarantee of anonymity. This possibility was confirmed in Sweden, where, following the passage of a law requiring the keeping of records of sperm donors which could be available to the children when they reach the age of eighteen, there was an immediate drop in the availability of AID. On the other hand, Rowland's survey of donors in Australia revealed that 42%

would continue to be donors even if they would no longer be anonymous (Rowland 1985). Of course, giving the child information about AID and even about the characteristics of the genetic father does not exclude the possibility of concealing the donor's specific identity. As George Annas of Boston University Law School points out, the interests of the child should prevail over those of the men who sell their semen (Annas 1979).

It appears that secrecy is also more for the protection of the husband than for the benefit of the child. This may even be recognized by the couple, who focus their concerns on what they judge to be the child's welfare. Protection of the nongenetic father is certainly one of the overriding concerns of the couples who use AID. The key difference between donor insemination and other methods is that AID is the only one which is used exclusively for male infertility or to avoid passing on a genetic abnormality in the male. Although the inability to have children is often assumed to be much more stressful for women than for men (Greil and Leitiko 1986), it appears that the couples, and society as a whole, consider male infertility a much more serious stigma than female infertility (Rowland 1985; Miall 1986). Men appear to regard infertility as more shameful. As one father said in explaining why he wanted his child's AID origins to be kept secret:

> I couldn't tell him. You know, I will have raised him all his life and I just wouldn't have the heart to tell him. I'm afraid he'd be ashamed of me. It might break my heart as well as his.

Support for the hypothesis that AID secrecy has to do mainly with protection of the social father comes from responses to questions about attitudes toward secrecy. Men are more likely than women to want to keep AID secret, and women are likely to cover up their husband's infertility by taking on the "blame" for reproductive problems.

Secrecy can be a difficult source of conflict within a relationship, since men are generally more committed to secrecy than women. This difference is especially hard on a woman who needs to talk but feels she must comply with her husband's wishes to keep their infertility problems a secret. In our research, most couples said they agree with each other about whether or not to tell others of their experiences. Some couples, however, particularly those who use AID, found this to be a problem. A woman whose baby was conceived with AID explained:

> My husband insisted we tell no one about the AID procedure, and that was very hard for me. I felt that, in trying AID, I would be living with a "lie" the rest of my life. I finally broke down and told a close friend. It felt like I was releasing an enormous pressure from my mind.

In most cases where there is a difference in opinion about infertility issues between husbands and wives, it is the husbands' wishes that prevail. Dr. Judith Lorber, a specialist in the study of health care and gender issues, writes that, in our society, men's wishes tend to win out in all areas of decision-making around fertility. For instance, she cites a study of childless couples in which it was found that if the husband wanted a child and the wife did not, they usually divorced; but if the wife wanted a child and the husband did not, they tended to stay together and not have children (Lorber 1985).

Not only do men's opinions usually prevail, but as one social worker remarked to us and as other observers have also noted (Berger 1980; Rowland 1985), many women said they would cover up for their husbands' infertility, saying that the problem was their own. If they had previously told friends that the infertility was their husband's problem, and then they used AID, they would change the scenario to protect their husbands. A woman who wrote to RESOLVE expressed her bitterness about having to cover up her husband's infertility by letting others think that she was the one with a problem:

> I was trying to get away from responding to infertility like a case of "cooties," something you feel compelled to pin on "the other guy." Even though I knew my sexual identity was intact, it felt like a hollow reassurance. I seemed to be the only one who knew this. . . . If everyone else sees you as infertile, it is hard not to react as though you are.

Another woman wrote to RESOLVE about the difficulties she was experiencing because her husband wanted no one to know that he had had a vasectomy during a previous marriage. The cover-up even affected the woman's relationship to her own mother, who felt guilty about her daughter's supposed infertility. Yet despite the obvious conflicts their cover-up presented, these woman and many more continue to pretend that they are infertile, continue to protect the man from what is apparently a greater social stigma.

If reproduction is supposed to be primarily a women's issue, why would infertility be more shameful, more stigmatized for males? There are a number of different approaches that may be taken to explaining this phenomenon.

A simple psychodynamic explanation would note that many men judge themselves and are judged by others by their ability to "perform," whether it be at work, in bed, or in producing offspring. A man who is lacking in any one of these domains feels deficient in all areas of life. This is particularly true when the problem is infertility since, in males, fertility is often confused with virility. One man who has a Ph.D. in biochemistry offered an explanation:

> Society has sterility and impotence all mixed up. Who should understand the difference between sterility and impotence better than I, but my first reaction to learning I was sterile was that I must be impotent. I should know better, but that was my first thought.

This idea can also be seen in another father's comments:

> When I told my son about AID origins, I also said that I still have an erection, so he would know that infertility is one thing and potency something else.

Infertile mothers rarely feel compelled to explain to their adopted children that they can have orgasms. The association of sterility with impotence or performance anxiety probably also explains why the great majority of couples who seek contraceptive sterilization choose tubal ligation over vasectomy, even though the former is a riskier procedure. It is another example of women being willing to subject themselves to a possibly unnecessary risk, whether physical or social, in order to protect a husband's concern about sexuality.

A second approach would consider the power factor in the couple's relationship to each other and to the child. AID represents an imbalance in the relationship of parents to child which does not exist with adoption or methods in which both parents contribute genetically to the conception (Rowland 1985). Not only is the mother genetically related, she has also experienced pregnancy and birth.

Male infertility may also disrupt the unspoken assumption of the man's dominance in a relationship, giving the woman more power than either of them feels comfortable with. Some wives even say they feel guilty about being "whole" when their husbands are not. There are accounts of women actually ceasing to ovulate when AID treatment begins. Do they feel, whether consciously or unconsciously, that they are supposed to be the ones who are infertile, especially since that is what everyone else assumes? Perhaps these women who take on the responsibility for infertility are attempting to restore the previous state of power in their families.

Finally, there is the explanation of men's need, in a patriarchal society, to demonstrate their biological relationship to their children as a means of maintaining their dominance in the family and in society. Donor insemination, as Corea (1985) points out, poses a threat to male dominance by putting patriarchal descent into question and making husbands less important to women for reproduction. "The Redundant Male" is the title of a recent book which reflects the concern of men that reproductive technologies such as AID will diminish their importance and control.

Conclusion

Male dominance in the family and in society is threatened by AID; the individual man's sense of himself and his general as well as sexual competence are also threatened. It is not surprising then, that most couples choose to keep the method a secret. This does create difficult personal dilemmas for many families, however.

The physicians who provide AID usually counsel silence; social scientists who have studied AID almost always encourage openness. Studies of adopted people suggest that the child's knowledge of his or her origins may indeed create problems, even turmoil. Yet they also show that a lack of complete information and understanding is even more harmful (Lifton 1979; Sorosky, Baran, and Pannor 1984). With the experts disagreeing and convincing arguments on both sides, parents who are committed to openness in their families yet fear the consequences of telling are faced with a harsh dilemma created by the benefits of technology. One man wrote to us describing his ambivalent feelings about this question:

> Ideally, I don't want to lie to my child or deceive her by failing to tell the whole truth. It doesn't seem right for me to decide that she doesn't need to know the truth about her conception. Yet, if I do tell her about it when she's old enough to understand, it could be too upsetting for her. After all, she would never be able to trace the donor if she wanted to. Why cause problems unnecessarily?

Jim is a father who resolved these questions by deciding to tell his son Michael, even though he worried about what the news would do to their relationship. He described to us his reasons for telling his son about AID and the "momentous day" when he did finally share with Michael the story of his conception:

> It was too stressful keeping this information from my boy. I don't think my wife and I ever sat down and said "We need to tell Michael. When should we tell him?" I think it was understood that it was my job, since I'm not the biological parent.
>
> I knew in his earlier years that he was too young to understand; he didn't have enough information to process what I was going to tell him. On the other hand, I wasn't going to wait until adolescence, because then, with whatever else was going on between us, this would just be thrown in the hopper. It would be a terrible betrayal. He's ten now, and I knew it would have to happen soon. I wasn't nervous because I hadn't planned it out; it happened really spontaneously, and I will just never forget it.
>
> I just felt so good after, I had not realized how much it took out of me to be keeping it from him. I felt like I had completed something, I had ended

> a travail. I really felt a burden lifted from me. It may be one thing to say they don't need to know, but it's another to say that you as a parent don't need to share it.

More parents are beginning to share Jim's view about AID. But the contrast with attitudes toward other methods is still very striking.

In order to test the hypothesis that male infertility is considered more socially unacceptable and that this explains the difference between AID and other methods in attitude toward secrecy, two other comparisons could be made. First, methods using female egg donors could be compared with those using male sperm donors, to see if there is a difference in attitude toward secrecy. Our interviews with people involved in the ovum transfer program suggests that the parents are more willing to talk about it with the children and with friends and relatives than are AID families, even though all of the conditions cited as important for AID secrecy also obtain.

Comparing methods which use donors would permit the examination of attitudes toward the importance of female vs. male genetic relationships to children. It may also reflect differing attitudes toward female vs. male adultery, since AID in particular has been described as a form of female adultery (Rubin 1965). It would also be worthwhile to consider methods which can be used either for male infertility or for female infertility but do not involve donors. This would apply both to IVF and to AIH. (IVF is increasingly being combined with both male and female donors, but we are suggesting that only non-donor cases be examined.) When we examined the responses of people who had used AIH and divided them according to which partner was affected, there was an obvious difference in attitude toward secrecy, with couples having male infertility problems more likely to keep the information secret. The numbers are very small, however, and the comparison would need to be carried out on a much larger scale. The evidence suggests that the issue of secrecy is a function of which partner is affected. Until male infertility is discussed more openly and its prevalence recognized, AID will continue to be veiled in secrecy and the families using it will continue to face difficult ethical choices and emotional dilemmas.

References

Annas, George. 1979. Fathers Anonymous: Beyond the Best Interests of the Sperm Donor. Genetics and the Law III, Aubrey Milunsky and George Annas, eds. New York: Plenum Press.

Beck, William, Jr., and Edward E. Wallach. 1981. When Therapy Fails—Artificial Insemination. Contemporary Obstetrics and Gynecology 17:113–125.

Berger, David. 1980. Couples' Reactions to Male Infertility and Donor Insemination. American Journal of Psychiatry 137:1047–1049.

Clamar, Aphrodite. 1980. Psychological Implications of Donor Insemination. American Journal of Psychoanalysis 40:173–177.

Corea, Genoveffa. 1985. The Mother Machine: Reproductive Technologies from Artificial Insemination to Artificial Wombs. New York: Harper and Row.

Curie-Cohen, M., L. Luttrell, and S. Shapiro. 1979. Current Practice of Artificial Insemination by Donor in the United States. New England Journal of Medicine 300:585–590.

Czyba, J. C., and Marie Chevret. 1979. Psychological Reactions of Couples to Artificial Insemination with Donor Sperm. International Journal of Fertility 24:240–245.

Greil, Arthur, and T. A. Leitiko. 1986. Couple Decision-Making Regarding Infertility. Paper presented to the Society for the Study of Social Problems, New York, August.

Lasker, Judith, and Susan Borg. 1987. In Search of Parenthood: Coping with Infertility and High-Tech Conception. Boston: Beacon Press.

Ledward, R. S., L. Crawford, and E. M. Symonds. 1979. Social Factors in Patients for Artificial Insemination by Donor. Journal of Biosocial Science 11:473–479.

Leeton, John, and June Blackwell. 1982. A Preliminary Psychosocial Follow-Up of Parents and Their Children Conceived by Artificial Insemination by Donor. Clinical Reproduction and Fertility 1:307–310.

Lifton, Betty Jean. 1979. Lost and Found: The Adoption Experience. New York: Dial Press.

Lorber, Judith. 1985. Gender Politics and In Vitro Fertilization Use. Paper presented at the Emergency Conference of the Feminist International Network on the New Reproductive Technologies, Sweden, July 3–8.

Manuel, Christine, Marie Chevret, and Jean-Claude Czyba. 1980. Handling of Secrecy by AID Couples. Human Artificial Insemination and Semen Preservation, G. David and W. Price, eds., pp. 419–429. New York: Plenum Press.

Miall, Charlene. 1986. The Stigma of Involuntary Childlessness. Social Problems 33:268–282.

Milsom, Ian, and Per Bergman. 1982. A Study of Parental Attitudes After Donor Insemination. Acta Obstetrica et Gynecologica Scandinavica 61:125–128.

Rowland, Robyn. 1985. The Social and Psychological Consequences of Secrecy in Artificial Insemination by Donor Programmes. Social Science and Medicine 21:391–396.

Rubin, Bernard. 1965. Psychological Aspects of Human Artificial Insemination. Archives of General Psychiatry 13:121–132.

Snowden, Robert, G. D. Mitchell, and E. M. Snowden. 1984. Artificial Reproduction: A Social Investigation. London: George Allen and Unwin.

Sorosky, Arthur, Annette Baran, and Reuben Pannor. 1984. The Adoption Triangle. New York: Doubleday.

Waltzer, Herbert. 1982. Psychological and Legal Aspects of Artificial Insemination: An Overview. American Journal of Psychotherapy 36:91–102.

10

Commercial Surrogacy: Social Issues Behind the Controversy

Linda M. Whiteford

Surrogacy has been described as baby selling, womb renting, an exploitation of women, and a miracle cure for infertile couples. The controversy surrounding surrogacy focuses attention on conflicting values, ambiguous social norms, changed social roles, and questionable legal standards. Surrogacy becomes a prism through which we can view social and, often contradictory, cultural meanings about family and parenthood. Our response to that refracted image reflects our cultural values. Those values, in turn, tell us about ourselves. The powerful emotional responses engendered by the Baby M case, in which the surrogate mother's rights to her child (Baby M) were pitted against the biological father's and social mother's rights to the same child, reflect the current concern and lack of social consensus about surrogacy. Analyzing surrogacy allows us to focus on the often implicit values attached to social roles within the family and to make them conscious, explicit, and available for discussion. In understanding the social values underlying critical elements of social organization, such as the family, we are better able to anticipate the consequences of change. Surrogacy, then, is prismatically reflecting, distorting, and refracting our cultural vision of the family—each facet throwing back an image slightly skewed from the one expected. That slippage, between the anticipated and reflected reality, focuses our attention on the contradictory values called into focus by surrogacy.

Commercial surrogacy refers to the contractual arrangement, usually mediated by an agency and involving an exchange of money, between a couple and a fertile woman. The fertile woman is paid to become impregnated through artificial insemination with the sperm of the husband. The fertile woman then becomes the "surrogate mother," carrying the fetus through pregnancy until its birth. At the child's birth, the surrogate, or biological, mother relinquishes the child to its biological father and its social mother.

The Surrogacy Process

In the United States, this process is frequently accomplished through the assistance of a third party or broker. The broker matches couples with women who are willing to act as their surrogates. The broker facilitates the selection of a surrogate, the artificial insemination, the medical care during pregnancy and birth, and the contract specifying rights and obligations of the parties involved. For those services, the broker receives a fee in addition to the costs for, and incurred by, the surrogate. The total cost may be between $20,000 and $30,000. One recently reported fee was $20,000 (Port 1987f); of the $20,000 paid by the couple, $10,000 went to the surrogate mother and the remaining $10,000 went to the broker. The couple paid the $20,000 and also all of the expenses, including doctors' fees, incurred during the process. Surrogacy is rarely covered by medical insurance; the entire cost must be borne by the couple desiring the child.

Although surrogacy contractual agreements may vary, there is similarity in the social process by which they occur. The following description, although fictitious, is based on published data and is designed to acquaint the reader with steps in the surrogacy process. It is a process in which no one is protected, in which all parties are vulnerable, and in which there are no guarantees.

Susan and Richard, both in their early thirties, have been married for ten years.[1] They delayed starting a family until they had finished school, repaid some of their financial indebtedness, and secured employment. Susan contracted an illness that resulted in a partial hysterectomy. Both Susan and Richard were despondent over the surgery, and, as time passed, Susan became increasingly depressed over her resultant infertility. Susan felt that her infertility denied them a chance to raise a child and condemned them to "half a life." Susan's depression grew into an obsession with pregnancy and children. During this time, Richard saw an article in a magazine describing the surrogacy process and brought it home to show Susan. They both decided to contact the agency described.

Fertility Finders (FF), the agency Susan and Richard contacted, told them that for a fee of $25,000 the agency could find a woman willing to be a surrogate for them. Susan and Richard were to sign a contract specifying that they would pay the required fee plus all of the medical expenses. Richard was to provide semen for the artificial insemination of the surrogate. In exchange, Susan and Richard could meet the surrogate, require medical fetal testing, and, if the tests showed fetal abnormalities, Susan and Richard could require that the surrogate abort the fetus. If the fetal test results were good and the fetus was carried to term, the surrogate would turn over the baby to Susan and Richard. They were delighted and decided to sign the contract with Fertility Finders and begin the surrogacy process.

Mary, who was to become the surrogate mother to Susan's and Richard's baby, was a twenty-four-year-old mother of two children. She was recently divorced and supplemented her child-support payments by working part-time at Sears. Since her divorce, Mary had felt lonely and isolated; when she saw the notice for surrogate mothers that Fertility Finders placed in a local newspaper, she decided to apply. Her own pregnancies had been easy, and she had felt happy and fulfilled while pregnant. The money mentioned in the advertisement was an added incentive. Mary contacted the agency, and FF asked her to come in for an interview. The people at the agency liked what they saw in Mary; she was young, had already established her fertility, seemed bright, and had no obvious physical deformities. After asking Mary some questions and having her fill out some forms, they told Mary that she was hired; they had a couple that would be a perfect match for her.

FF was slightly concerned that Mary was divorced. They felt that her sexual activity would be more difficult to control if there was no constant male in her home. Mary's marital status, however, made it easier for Richard and Susan to be able to adopt the child. In many areas, state laws determine that if a surrogate mother is married, her husband automatically becomes the legal father of the child; however, because Mary was divorced, Richard could be listed as the father and Susan could adopt the child.

Susan and Richard agreed to pay the $25,000 fee, in addition to Mary's medical expenses, and to provide viable semen. Mary agreed to be inseminated; to refrain from drinking alcoholic beverages, smoking, or taking unprescribed drugs; to abstain from intercourse from two weeks before insemination until pregnancy was confirmed; and to keep all scheduled administrative, medical, psychological counseling, and legal appointments arranged by FF for her. All medical appointments were to be made with providers selected by and approved of by the agency. Mary also agreed to submit to all medical procedures required by the agency's physicians.

Mary was promised $200 for the purchase of maternity clothes and fifteen cents for each mile she traveled to meet required medical appointments. She understood that if she did not become pregnant, she would not receive any money from the agency, nor would she receive any payment if she suffered a miscarriage. If the child was stillborn, born with deformities, or if the father refused the child, Mary would be paid her $10,000 only after the agency physician determined that Mary was not at fault.

Mary became pregnant. She got to know Susan and Richard, who both played active roles in Mary's life during her pregnancy, labor, and the birth. Susan and Richard wanted to be part of the process that was creating their child, and they established a friendship with Mary. They made sure that Mary took her prenatal vitamins and encouraged her to eat well and get exercise. They played with Mary's two children. Susan and Richard insisted, even though Mary objected, to amniocentesis for a prenatal diagnosis of the fetus. At the time of birth, Mary relinquished the baby to Susan and Richard and the relationship with the couple ceased. Now Susan and Richard were joyous; they had their own son. Mary, with a mixture of happiness in her role of providing Susan and Richard with the source of their joy and sadness at losing the child who was her companion of the last nine months, accepted her payment and returned to her children.

The couple and the surrogate mother were asked after the successful birth if they would do it again. Susan and Richard said they would do it again if they could afford the process. Mary, on the other hand, said that she would not. The money was good, but she would not want to experience again the pain that accompanied the loss of her baby and the termination of her friendship with Susan and Richard. Her life now seemed even more lonely, and she felt more isolated than she had before the pregnancy.

The four primary participants—the couple, the surrogate, and the broker—each have different motivations, anticipations, and concerns for entering into, or continuing with, surrogacy arrangements. The broker will continue to provide surrogacy services as long as the process is socially sanctioned, there is a demand for the product, and there are enough women willing to be surrogates. Couples like Susan and Richard who strongly desire a child genetically related to themselves and who can afford the $25,000 to $30,000 will continue to go to agencies who will mediate the surrogacy process, especially if they feel protected by an enforceable legal contract. And, as long as women are financially and emotionally vulnerable, there will be surrogates.

Consequences of Commercialization

Surrogacy is not new; women have borne children for infertile women before. In the past, surrogacy occurred between sisters or within families, was rarely reported, and money was seldom exchanged. The current situation varies from that of the past in several significant ways: the use of a third-party mediator, exchange of money, occurrence of the arrangement between strangers, and basing the procedure on a quasi-legal contractual agreement. Each of those variables affects the nature and quality of the relationship between the surrogate and the couple.

The commercial use of women to act as surrogates during pregnancy formally separates the role of mother into two categories: biological mother, or surrogate, and social mother. In most societies women raise the children they bear: The biological and social mother are the same person. Less often, a woman bears a child who is then raised by some other person, often a relative of the mother. In such situations the biological and social mothers are two different people. In the United States the practice of commercial surrogacy enforces the distinction between surrogate and social mother through the use of an extrafamilial agency to broker the pregnancy. The brokering agency arranges the payment, usually occurring between strangers. The exchange of money between strangers further separates the women who bear children conceived by contract from the women who raise those children.

The practice of planned production of a child in exchange for money, in which conception occurs without sexual intercourse and outside of marriage, challenges contemporary social values. These values include the importance of continuing a bloodline, the sanctity of the family, the child-parent bond, and the importance of children. The potential effect of commercial surrogacy on the social roles that constitute the family is the focus of this chapter. For purposes of discussion, this chapter is organized around the social, ethical, and legal issues generated by commercial surrogacy and the social roles they affect.

The controversy over surrogacy turns on three major issues: (1) social issues, including differential power, class exploitation, and a redefinition of social roles of parenthood; (2) ethical issues, including rights, duties, and obligations of the participants; and (3) legal concerns, including the status of surrogacy agreements and the redefinition of maternal rights. Although each major issue is intrinsically interwoven with the others (enactment of laws that sanction moral decisions create and destroy social definitions of roles and obligations), each will be analyzed separately.

Commercial surrogacy exploits socioeconomic class differences, using financial need and emotional need as currency. The exchange of money transforms surrogacy from an altruistic gift between sisters or friends into baby selling or womb renting and powerfully affects social relationships. The exchange of money for surrogacy codifies power and class inequities between those who can afford to buy new life and those who sell their ability to create life. The exchange of money for surrogacy categorizes people as buyers and sellers, categories based on socioeconomic differences.

The role of the broker or third-party mediator further removes personal involvement and obligation from current surrogacy practices. Commercial surrogacy emphasizes two primary characteristics that distinguish the purchasers from the provider: money and fertility. It is clear that each party to the surrogacy agreement must have something that the other desires. The resultant exchange provides a child for money; an exchange founded on differential access to economic resources.

There is no clear consensus of opinion about surrogacy. Neither the public nor those people whose professions or politics involve them in the issues of reproductive technology show unilateral agreement on the subject. Physicians and ethicists disagree among themselves over the issues generated by the discussion of surrogacy. Feminists also find surrogacy a complicated issue to resolve. During the court case of Baby M, some feminists supported the surrogate mother's right to have access to her child (and her right to abrogate the surrogacy contract), while others supported the right of women to choose surrogacy (thereby supporting the surrogacy contract). The lack of consensus reflects the complexity of the issue and the concomitant necessity for careful analysis. Ann Snitow, who referred to herself as a "feminist lost in the thicket of new reproductive technologies," provided an insightful comment on the need to establish guidelines that neither reestablish "the myth of the eternal mother with her unending maternal responsibility" nor deny the validity of women's desires to be mothers (1986:77). "The new reproductive technologies cause unease because they insert a scalpel between mother and father, parent and child, egg and sperm, X chromosome and Y" (Snitow 1986:77).

Social Issues

Why does surrogacy raise such contradictory feelings in people? Is surrogacy, as some have claimed, a threat to society (Snowden, Mitchell, and Snowden 1984)? Will it change the structure of the family, of patterns of inheritance, of basic relations between men and women? Snowden, Mitchell, and Snowden argue that surrogacy does threaten the structure of the family and thereby threatens the stability of society. This threat occurs because surrogacy removes social controls over reproduction by separating reproduction from sexual intercourse. Because reproduction no longer needs to be preceded by sexual intercourse, it can no longer be controlled by legal and cultural rules concerning who can have sexual intercourse with whom. Social controls aimed at limiting legitimate sexual access through incest taboos, age of consent, and marriage rules no longer determine reproduction. AID takes the private and personal activity of reproduction out of the realm of physical and emotional connections and makes it a public commercial transaction mediated by professional brokers—namely, lawyers, doctors, and donors.

Families do form an integral part of society; they are the basic unit of enculturation, the cement of the kinship system. The birth of a child is an event of social significance, important because of the individuals and families who are connected genetically and nominally through that child. People are connected to the generation immediately preceding them, not only by the values shared intergenerationally, but also by the physical attributes that are passed on. It is, however, an ethnocentric view to assume that any single form of the family is crucial to the survival of any society. From a cross-cultural perspective, the family is an unusually dynamic and variable element. Societies exist in which biological parenthood is not recognized to be distinctive from social parenthood (Goodenough 1970; Fox 1968). Most societies do, however, recognize the woman who bears the child to be the child's mother. In the case of surrogacy, the roles are reversed. The woman who bears the child is not socially recognized as being the child's mother, and a man to whom the woman is not married and with whom she has not had sexual intercourse is socially recognized as being the child's father.

A recent edition of a national newsmagazine ran a cover feature on surrogacy in which the stereotypes were detailed (Lacayo 1987). One photo showed a couple in front of their large, new home with their two surrogate children; they had already had children but wanted more children, and because the wife was postmenopausal, they chose surrogacy. Another photo showed a happy surrogate mother who said that she loved being pregnant and chose to be a surrogate because she wanted to make other people happy by "giving" them a child they could not otherwise have. She was willing to do this for others even though during her previous pregnancy she had been hospitalized twice because of severe morning sickness.

The media presents contradictory stereotypes of the people involved in surrogacy. The surrogate mother described by the newsmagazine is pictured as a warm, loving, and altruistic person, but also as someone who appears irrational or illogical, as she willingly enters into a state that on previous occasions sufficiently threatened her health and that of her unborn child so that she twice required hospitalization. A readers' poll about surrogacy conducted by a major newspaper generated a response by 1,600 readers (Mason 1987). One of the questions asked of readers was whether or not surrogacy should be allowed. The readers' responses were split; 51 percent of them said yes, surrogacy should be allowed, whereas 49 percent felt that surrogacy contracts should be made illegal. The split in public opinion shown in this poll reflects society's indecision about surrogacy.

The surrogate mother and her family, including her husband and their children; the children who are born through surrogacy; the biological/social father of these children; and their social mother are the primary actors in the surrogacy drama. In the following discussion, I apply a social and ethical analytical framework to the analysis of current information about surrogacy, adoption, and familial interrelationships in order to raise questions that focus attention on the personal consequences of surrogacy.

Surrogate Mother and Her Family

At least one study is presently being conducted on surrogate mothers in an attempt to find out who they are and why they are willing to conceive and carry a child that they plan then to give up (Parker 1983). Controversy has arisen over what these women do and why they participate in surrogacy programs. They have been portrayed as prostitutes, "breeders" working for "stables," exploited women, angels from heaven, and poor women working a job (Corea 1985). Some surrogate mothers say that they participate in surrogacy programs because they enjoy being pregnant and they want to make other, childless couples happy. In addition, they need the money surrogacy brings (Martin 1976). Parker (1983) also suggests that some women are willing to be surrogates as therapy for unresolved feelings experienced after a previous abortion.

Dworkin, on the other hand, points out that surrogate mothers sell their genetic material and lease their bodies in a male-dominated political and social system. She comments, "In both prostitution and surrogate motherhood . . . the state has constructed the social, economic and political situation in which the sale of some sexual or reproductive capacity is necessary for the survival of the woman. It fixes her social place so that her sex and her reproductive capacity are commodities" (1987:228). Young, poorly educated women with few job skills have limited options in current job markets. Surrogacy offers an economic alternative. It does not, however, prepare the surrogate mother for enhanced future job opportunities. Surrogate mothers are removed from the public job market during their pregnancy, thereby making it more difficult to find other employment after the pregnancy is completed. Because of the constraints on the surrogate's time stipulated in the contract, such as numerous prenatal appointments with physicians, a forty-hour-a-week job is difficult to maintain during a surrogate pregnancy.

By signing the surrogacy contract the woman also makes herself vulnerable to medical intervention, medical procedures decided upon by others who do not necessarily have her best interests in mind. A situation could occur in which the interests of the surrogate mother and those of the social parents were in conflict. According to the contract described earlier, the social parents' interests would be honored over those of the surrogate mother. The surrogate mother might not want prenatal testing, fetal monitoring, or a cesarean section, but the social parents could insist that those procedures be carried out. Some surrogacy contracts specify not only that the surrogate mother must undergo prenatal tests such as amniocentisis but also that if the biological father insists, the surrogate mother must abort the fetus. Such a contract is a violation of a woman's constitutional right to privacy over the decision of abortion. The right to privacy concerning abortion has been upheld by the Supreme Court of the United States and, as a constitutional right, cannot be relinquished by a surrogate mother or anyone else on her behalf. Yet, its relinquishment is included in some surrogacy contracts.

Not only do contracts often fail to protect the surrogate mother from medical intervention and from potential conflict-of-interest disputes, they also do not assure her the right to confidentiality. Even when contracts make that promise, brokers may not be able to substantiate it, as the surrogate mother's name appears on the child's birth certificate, thereby providing her identity to anyone wishing to find her. Serious questions have been raised about the ability of surrogacy brokers, such as some infertility clinics, to fulfill obligations of protection of confidentiality.

The right to informed consent is also called into question by surrogacy contracts. Informed consent is designed to protect the unwary and the uninformed. It can be assured only after a minimum of three conditions are met: full knowledge of potential consequences of the activity, determination of mental and emotional competency of the subject, and unintimidated choice (Freedman 1975). Informed consent states that clients are told of potential complications and risks consequential to the procedures they are to undergo. It may not be possible to inform a surrogate of the full potential of psychological and emotional trauma she could experience when required to relinquish the child (Singer 1985; Bayles 1984). A recent study showed that one-third of the population of surrogate mothers contacted required postpartum counseling because they were so distraught after having to give up their babies (Goleman 1987). Surrogacy contractors may meet the legal guidelines required for informed consent, but it may be impossible to comply with the ethical guidelines of informed consent when applied to the emotional and biological aspects of surrogacy.

Lack of government regulation and a desire for privacy has, until the case of Baby M, kept surrogacy from full public view. The popular press focused attention on the positive aspects of surrogacy between 1980 and 1986 but paid little attention to the possible negative implications of such arrangements. Feminist scholars such as Dworkin and Corea have pointed out the negative consequences of surrogacy for women, but few writers have looked at the potential consequences of surrogacy on other members of the family. As a result of current litigation, expert witnesses have presented testimony illuminating potential emotional crises resulting from surrogacy. Family members of surrogates, such as husbands and children who are already members of the family, are also affected by the process.

Snowden, Mitchell, and Snowden (1984) suggests that children already born to a surrogate mother could experience terror associated with the fear of being given away. Childhood insecurities could be magnified and relations with the mother could be adversely affected. One surrogate mother said that after she gave up her AID baby, her other child asked: "If I am bad, will you give me away, too?" Surrogacy intervenes in the lives of the surrogate's family in ways not yet fully known.

Presently there is no standardized accountability for a child conceived through surrogacy. In some states the surrogate's husband is legally responsible for children born to his wife during his marriage to her. According to laws in these states, that same liability extends to all children born during the time of the marriage, whether or not their conception occurred through artificial insemination. Legal ambiguity concerning responsibility for an AID child may place the surrogate's husband in a dilemma. Even if he is aware of the surrogacy arrangement and signs a contract to that affect, he may be held legally liable as the "father" of his wife's child. If he does not officially acknowledge the artificial insemination origin of the pregnancy, he may void the contract between his wife and the agency and still be viewed as legally responsible for the child.

Theoretically, the husband of the surrogate should not be responsible for an AID-produced child. His legal responsibility for the child is ambiguous. His emotional attachment to the developing fetus is unknown, though both the duration and public visibility of pregnancy would make it difficult for him to ignore.

Children Produced Through Surrogacy

The ambiguous social acceptability of surrogacy has made it difficult to study the development of the children born through the process, although the article by Lasker and Borg in this volume (Chapter 9) provides crucial insight. Are the children born through surrogacy told they were adopted, that they were the genetic creation of both parents, or that they were born to a surrogate mother? Research on adoption has shown that what a child is told about his or her birth, how that information is given to the child, and when the child learns about these birth circumstances have a powerful influence on the child's sense of self. Parents who are uncomfortable with the social and ethical implications of surrogacy, who perhaps do not wish to acknowledge the role played by the surrogate mother, cannot be expected to be comfortable telling their child the circumstances surrounding his or her birth. It is possible that denial of surrogacy could undermine parent-child trust.

Perhaps children born through surrogacy bond twice, once prenatally with their surrogate mothers and again, a second time, with their social parents. In this, children born through surrogacy are similar to adopted children. Will the children of surrogacy, like some of those given up for adoption, also search for their "birth mothers?" How might a child feel to learn that he or she was born as part of a contract? Would that knowledge influence an individual's perception of himself or herself? In the complex human set of emotions played out within the dynamics of the family, it is impossible to predict individual reactions. Michael Bayles, in his book *Reproductive Ethics* (1984), writes that children born through surrogacy might find it disturbing to learn that they were given up and might, as some adopted children do, feel that their mothers rejected them because of some personal flaw in the children.

Expert witnesses for the defendant in the Baby M custody trial reported that adopted children, in general, have lower self-esteem than do nonadopted children (Port 1987g). One such witness suggested that children born through surrogacy might believe the world a "hostile" place and could be at serious risk of psychological problems. If the surrogate mother completes her contract and relinquishes the child, the child might later ask: "Why did she sell me?"

Genetic Father/Social Father

In the case of AID surrogacy, the genetic father is also the social father. Little is known about the men who become fathers through surrogacy or what effect surrogacy has on their lives. In surrogacy, the father is defined by his reproductive and legal role. He contributes his genetic material to the child, and it is he who, as is stated in the surrogacy contract, has the right to refuse to accept the child (Ince 1984). Interviews with surrogate mothers reveal that although they refer to "giving the child to the couple," it is clear that the father's genetic contribution strongly shapes their perception of his role. One surrogate conceded that childless couples could adopt an older child, a child of another race or nationality, or a handicapped child, but that "these people want a child related to them, who will carry on their [i.e., the husband's] bloodline" (Corea 1984:22).

Would the experience of fathering a child born through a surrogate differ from the experience of fathering a child resulting from sexual intercourse with the child's mother? The financial and emotional costs during pregnancy are different for the father of a surrogate-born child than for fathers of other children. Because extreme action has been taken to make possible the creation of the surrogate-carried child, the father may have a high degree of ego investment in the child. The relationship between the father and his wife may change if the father feels greater responsibility for the child because of his genetic relationship to it. The genetic/social father is in the unique position of having to adopt his own child because, although the child carries the father's genetic material, the child at birth may be legally considered the child of the surrogate mother and her husband.

The relationship between a husband and wife changes after the birth of any child. Would commercial surrogacy place barriers between the genetic father and the social mother because they are differently related to the child? At the moment, little is known about how surrogacy might affect the husband-and-wife pair, but documentation is available concerning couples who have chosen to have the wife artificially inseminated with donor semen (Snowden, Mitchell, and Snowden 1984). Cases in which the biological and social father are not the same man, social fatherhood is similar to, but not the equivalent of, social motherhood through surrogacy. AID fatherhood, like social motherhood through surrogacy, is a legally recognized relationship between a nongenetically related parent and child. AID fatherhood (in which the wife is artificially inseminated) and social motherhood through surrogacy (in which the husband's sperm is used to impregnate a surrogate) share a characteristic lack of contribution of genetic material to the child. However, they differ in their outward timing and involvement with the pregnancy. AID fatherhood can begin the moment the wife becomes pregnant. The social father is present, socially acknowledged, and takes an active part in his wife's pregnancy and the birth of the child. His wife carries and gives birth to the child, thereby making her husband the father of the child in the eyes of the community.

In the case of AID surrogacy, the genetic father is not the husband of the woman pregnant with his child and is not socially acknowledged to be the child's father until after the child's birth. In commercial surrogacy, there is always the fear that the surrogate mother may be unable to relinquish her child to its genetic father and his wife. That fear and the lack of social recognition of his role may affect the genetic/social father.

The Social Mother

The person mentioned least in the literature on surrogacy is the social mother. She also has few rights in the legal documents and contracts, yet she will play a major role in raising the child. Her social role is described by her biology: She is the infertile wife. A secondary, frequently posited characteristic of the social mother is her altruism: She wants her husband to be able to reproduce his own genetic material, even if she is not involved in the procreative process. Granting that different women who are social mothers to children born through surrogacy have individual characteristics, some characteristics shared by members of a society might be hypothesized.

Infertility in a pronatalist society often engenders anxiety and guilt. For this reason, AID practices tend to encourage secrecy, and both infertility and AID have a social stigma associated with them (Crowe 1985). A woman who grows up in a society in which she sees marriage and raising a family as her main function, and who marries a man who shares that view, may respond to her infertility with depression and guilt. If, to ameliorate that pain, she fulfills social expectations and allows her husband to pass along his genetic material through a surrogacy arrangement, what are the implications for her? What kinds of responses to the surrogacy experience might she encounter?

Although she may feel pleasure in anticipating the birth of her husband's baby, she may also feel anxiety about her own role as mother. Anger, jealousy, and guilt may also influence her feelings. She may feel a sense of guilt in regard to her own infertility. She may be jealous of an arrangement whereby another woman carries the baby who embodies her husband's genetic material but not hers. The social mother might be angry over her own infertility and resent the surrogate mother who is so fertile she can give birth to a baby and give it up. She might be hurt that her husband's need to have his biological child has been fulfilled at the expense of her own exclusion from the process.

Having a child is a long-term commitment that continues throughout life. The longevity of the child-parent bond and the responsibility that it entails create special problems. If a couple divorces, who claims the surrogate-born child? Could the social mother, who has no biological relationship to the child, refuse to claim the child? Is there a constant disequilibrium created between the husband and the wife because the child carries his genetic material but not hers? Might the social mother feel that she is raising a stepchild with all of the attendant difficulties of such alliances, or does she consider the child her own and, thus, see herself as a primary decision-maker?

In a retrospective study conducted in England of 899 couples who received artificial insemination by sperm donor between 1940 and 1980, responsibility for the resultant child was felt most strongly by the parent who contributed genetic material to the child's creation (Snowden, Mitchell, and Snowden 1984:90–91). The retrospective study included no cases of maternal surrogacy; all couples were composed of fertile women whose husbands were infertile. Women who became pregnant through AID felt more responsibility for the child, because of the process by which they became pregnant, than did their husbands. When the child was disruptive, women said they tried to protect their husbands from the disturbance. In the same study, one husband discussed his emotional response to his lack of genetic connection with his AID child by saying, "I suppose that it's the point of concern . . . that when you get to adolescence and the balloon goes up as it were, it's easy to shelve off responsibility by perhaps even in anger actually saying: 'Well he isn't mine anyway. This is nothing to do with me.' And it does kind of worry me that one hopes to have the equanimity and stability to weather any storms of that kind" (Snowden, Mitchell, and Snowden 1984:97).

In AID surrogacy, the social mother has no role in the conception, pregnancy, or birth of the child who is fathered biologically by her husband. The social mother is not officially part of the family group until after the surrogate has given birth to the baby and has relinquished the child to its father. The baby is conceived, carried, and borne by another woman, and this process does not include the social mother until its completion in the actual birth. Like the genetic/social father, the social mother has no guarantee that after the baby is born it will even be given to her to help raise.

The Role of the State

Concern about surrogacy is expressed by lawyers (Dickens 1984; Garcia 1986), psychologists (Parker 1982), bioethicists (Annas 1979, 1986a; Juva 1985;

Singer 1985), feminists (Corea 1985; Arditti 1984; Dworkin 1984), and other social scientists (Snowden, Mitchell, and Snowden 1984; Whiteford n.d.). Although some observers want the state to regulate the process of surrogacy to protect the various parties involved, others believe the process should be outlawed. State involvement with surrogacy raises legal, social, and ethical issues, such as the rights and obligations of various parties, the legitimation of practices, and intervention in reproductive rights. The state functions to protect those elements important for its survival, stability, and continuity. The family, as a procreative and enculturative unit, supports the state and is regulated by the state. It has been argued that the state has an obligation to protect the vulnerable from harm and that in the legitimation of surrogacy the state fails to protect the unborn child and the family from the harm that is caused by unpredictability and the potential loss of trust (Snowden, Mitchell, and Snowden 1984:148).

Ethical Issues

Perhaps the most difficult aspect of surrogacy is that it takes us into an ethical unknown, a perilous trip for which there are few guides. We cannot refuse to explore the unknown in fear of where it might lead us, yet we must explore with caution and with patience. Analysis of surrogacy, by its very mix of the known, the unknown, and the unpredictable, forces the unraveling of a conundrum.

Is surrogacy a way for childless couples to have children and pass along genetic material, or is it a way to manipulate for profit a socially disadvantaged group of people? Will surrogacy result in their "ideal" families becoming realities for couples, or will it result in the creation of an underclass of women who are "breeders" in surrogacy stables where babies are made and sold for a profit? Will surrogacy provide social parents the opportunity to select certain physical characteristics such as eye color, stature, hair color, or will it create children who always wonder who their "real" mother is? Does surrogacy introduce possibilities of security and stability into a family by providing the social acceptability of having children, or does it introduce the influence of distrust and unpredictability into the family unit? If the ideal family is a place of personal security and support, as well as an integral part of the social system, then the values of trust and truthfulness that form the basis for predictability of behavior are important.

An English social researcher argues that surrogacy undermines family cohesion and unity by introducing new elements of uncertainty and unreliability. The researcher states, "We expect the bus to travel to the destination marked on the front and the cost of petrol to be that shown on the meter. We believe people when they provide information from an authoritative position and we feel angry and frustrated if any of this trust is shown to be misplaced" (Snowden, Mitchell, and Snowden 1984:148). The stigma associated with infertility and the ambiguous social acceptance of surrogacy may cause parents to falsely represent the biological origin of a child conceived through AID and born through surrogacy. This conspiracy to hide the truth may well have hidden personal and social costs for both children and parents.

The ethical quandaries posed by surrogacy require critical reflection. It is important to distinguish the elements that differentiate surrogacy from other forms of AID. The contribution of genetic material and a womb by a woman who gives the child up after its birth in exchange for money are the unique characteristics of surrogacy that are most controversial. Commercial surrogacy not only isolates sexual intercourse from reproduction, it also isolates the biological-social father from the social mother and each from the biological or surrogate mother. This isolation allows for the objectification of children as commodities, women as breeders or nurturers, and men as users. It encourages the belief that anything, including progeny, may be purchased.

In a capitalist economic system, with decentralized medical regulations and a powerful, monopolistic medical hierarchy, surrogacy has implications for the status of women both in the United States and in other countries (Fee 1983; Navarro 1977). Supporters of surrogacy have suggested that once the procedure is more widely legitimated, the cost to consumers, the social parents, will decrease, thus making the product—the baby—accessible to more potential consumers. The same people suggest that in the future surrogate mothers might be sought from countries in the developing world, where women could be contracted to bear children, using the money to support themselves (Corea 1985:245). In this future view, as one broker suggested, "the woman would be paid nothing for her services. The couple adopting the child would provide the surrogate's travel and living expenses. Though such women receive no pay . . . they benefit from the arrangement because they get money [on which to live]" (Personal Communication 1986).

Surrogacy is not a cure for infertility; it is the treatment of a social, not a physical, need. The practice of surrogacy has social implications and is a reflection of social values. It raises questions about the values of a society that creates a group of people who have no reproductive rights, whose bodies are subject to medical interventions not necessarily in their best interests. Such groups have been created in the past, wet nurses and slaves among them. Is commercial surrogacy opening the way to exploitation of a social class by using vulnerable women to provide a service to those more materially advantaged than themselves?

All of the AID forms of reproduction affect the social control of reproduction. Those who can pay have access to services, those who cannot pay either have no access to services or become part of the services by selling themselves. It is impossible to see into the future, but it is not difficult to postulate that surrogacy raises complex and disturbing questions. Surrogacy is a problem that cannot be avoided by claiming ignorance; we are already engaged in the process of creating "special-order" children. Resolving the legal issues that surrogacy generates will not solve the ethical issues that are embedded in the problem. From an anthropological point of view, no single and unique form of the family is necessary for human survival. From that same point of view, the sociocultural implications of surrogacy suggest possibilities for commercial use of men, women, and children involved in this form of future family structuring.

To obviate some of those possibilities, the conclusion reached by a team of researchers in the United Kingdom was: "The carrying of a child, following artificial reproduction, should be disallowed where the carrying mother is not also to be responsible for the nurture of the child" (Snowden, Mitchell, and Snowden 1984:171). West Germany is presently considering a ban on all surrogate activities in which the genetic or carrying mother is not the same as the nurturing or social mother. Australia and Israel and the United Kingdom have already outlawed commercial surrogacy.

Surrogacy by contractual arrangement between the natural father and social mother and the biological mother raises complicated issues of rights, duties, and obligations. Whose rights should be protected—those of the unborn child; the child's biological mother; the child's biological father; the contractual, or social, mother? Whose rights take precedence over those of the biological mother and her unborn child? Should the use of force to reduce risk to the unborn child be sanctioned? Is the biological mother only a "vessel" used to produce a child? What are the mother's rights, does she have a right to control her own body and to keep her own child? What are the ethical implications of a contract that, even before conception occurs, denies either the biological mother or the biological father access to the child? How is it possible to "do no harm" in such a situation?

Situations previously taken from the realm of science fiction now are real: extracorporeal fertilization, fetal surgery, the possibilities of posthumus procreation, genetic manipulation. Possibilities that were once barely imaginable now exist, and they raise new questions about the nature of protection and conflict of interest. As the developing fetus is provided with rights legitimated by the courts, the private, symbiotic relationship between a pregnant woman and her fetus holds the potential of becoming the focus of public, court-ordered actions. A woman might, for instance, be ordered to undergo a court-ordered cesarean section, or she might be held accountable for damage done to the fetus during her pregnancy.

What rights does the biological father have? Does his duty to the child carry the same obligatory power if that child is born deformed? In some surrogacy contracts it does not. The surrogate mother may be obliged either to abort the fetus or raise the deformed child herself, as required by the contract. Even without having to face those actual situations, acknowledging their possibility forces us also to consider the rights of other children in the surrogate's family. What are their rights? What values do they learn in this process? What are the rights of children born by commercial surrogacy?

Even terms of reference such as "mother" become confusing in cases of surrogacy (Snitow 1986). It has been argued that both the wife and the pregnant woman are surrogate mothers. Each woman contributes to the continued development of the child. The pregnant woman contributes her own genetic material and carries the fetus until its birth; the wife nurtures the child after its birth. Yet, neither woman does both. The husband who contributes both his genetic material and his time raising the child is both the biological and social father even though the child is conceived outside of marriage and without sexual intercourse. At the moment we do not know if these variations on traditional methods of procreation will have long-lasting effects on those children and their parents, but commercial surrogacy does force us to look carefully at culturally engendered assumptions about the meaning of parenthood and the family.

A child conceived and born through surrogacy is legitimated by an adoption process, the legal acknowledgment of the couple's social role. Indeed, in most cases, both the father and the social, but not biological, mother adopt the child born through surrogacy and thus formally and legally affirm the child as a part of their family. Snowden and Mitchell (1983) has argued that this form of adoption differs from the traditional form of adoption, in which there is no actual planning of the pregnancy. With the surrogate technique, a child is deliberately created for adoption. Although society has developed an acceptable process for traditional adoption, it has no socially accepted precedent for adoption of a child produced through surrogacy. In adoption cases, the birth mother has a five-day "cooling-off" period after the birth of her child during which she can, without penalty, reverse her decision to relinquish the child for adoption. Surrogacy contracts at present provide no such opportunity.

In the case of AID surrogacy, it is clear that medical technology has developed faster than the social structure of which it is a part. Medical technology is making possible changes in human behavior for which there are no names, no legal structure, and no social infrastructure. The implications of these changes cannot be fully anticipated; often, because they have no referents, they can only be discussed by analogy. Surrogacy is an example of how interventions made accessible through breakthroughs in medical technology have consequences that conflict with social values.

Legal Issues

Much of the literature on surrogacy focuses on the rights and obligations of the various parties and delineates ways of insuring that the parties involved will uphold their contractual obligations (Corea 1985). Too often the primary focus in such contracts rests on the father and his rights and, to a somewhat lesser extent, on the surrogate mother and her obligations. Little attention is paid to the rights of the child or to those of the social mother. A review of the literature points to the following legal issues:

1. determination of whether surrogacy is "baby selling," or the selling of a service (Corea 1985)
2. "quality control" issues concerning the screening of donors, surrogates, and appropriate application of medical technology (Poland 1986)
3. determination of paternity and legitimacy of offspring created through AID surrogacy (Dickens 1984)
4. contractual obligations between donor and surrogate (Garcia 1986)
5. exploitation of women and the creation of an "underclass" for reproductive use (Dworkin 1984)
6. issues of reproductive privacy (Arditti 1984)
7. role of the state in the protection of its people (Dickens 1984)
8. equality under the law and rights of mothers to their children (Chessler 1987)

At present, there is no uniform code of contractual arrangements for commercial surrogacy. In 1986, Baby M's birth brought about a contested court case. Only two such contracts had previously come to the attention of the courts. In Indiana, a judge ruled against the validity of the surrogacy contract; in Kentucky, another judge ruled to uphold a surrogacy contract (Port 1987e). The lack of consistency in court rulings may reflect a social ambiguity concerning the process. Garcia (see Chapter 11) suggests that new laws need to be established in the area of surrogacy. She points out the following areas of legal confusion that need clarification:

1. Under what, if any circumstances, may a buyer (infertile couple) refuse to accept the baby because of a patent defect?
2. How long after the initial "inspection" does the buying couple have to revoke acceptance of the child upon discovering a latent defect?
3. Who will take the baby upon the death or divorce of the infertile couple before the baby is delivered?
4. What liability rests with the surrogate mother for genetic defects that she passed on to the child if she knew, or should have known, that

she carried the genetic material that resulted in the defect and caused the infertile couple to reject the child?

5. What responsibility does the state bear to accept and care for children who are rejected by parties to surrogate contracts?

These issues explicitly involve the state in the dilemma over reproductive rights. The rights critical for consideration are often in conflict with one another. Do all couples have the "right" to have children? Under what circumstances may an infertile couple hire someone to "aid" them in achieving conception? What are the rights of the surrogate mother in regard to the child she conceives and carries? What rights does the unborn child have, and who will be granted guardianship for those rights? The state plays an important role in these decisions through its formulation and legitimation of the surrogacy process.

If the courts uniformly rule that surrogacy contracts are invalid on the grounds that they constitute baby selling, or that they do not demonstrate adequate protection for informed consent, the number of couples choosing surrogacy would decrease. It has been argued, however, that if commercial surrogacy is made illegal, as it is in some parts of the world, the practice would not disappear but would only go underground. The alternative, to legitimize surrogacy, requires the codification of adequate means to protect the rights of individuals whose interests, by definition, are in conflict with one another.

The contractual agreements pertaining to surrogacy raise a second set of moral, ethical, and legal issues. No federal court has yet determined the legal status of surrogacy contracts. At present, twenty-eight states in the United States have laws covering artificial insemination in which the consenting husband of the inseminated woman is considered to be the legal father of the child (Lacayo 1987). Surrogacy requires just the opposite, that the husband of the inseminated woman not be considered the father of the child, but that the donor be recognized as the legal father of the child. Even those statutes that cover artificial insemination seem to contradict the social requirements for AID surrogacy.

The legal status of surrogacy contracts is still being tested. According to Bernard Dickens, a professor on the faculty of the University of Toronto School of Law, the status of surrogacy agreements as legally enforceable contracts is very doubtful (Dickens 1984). Both Dickens and Garcia (see Chapter 11) suggest that no mechanism exists by which to extract remedy or repayment if the surrogate mother breaks her agreement and keeps the child. That is, based on the legal status of the surrogacy agreement alone, the surrogate would most likely be allowed to keep her child. However, when the powerful economic differences that are at the very heart of commercial surrogacy are considered, the outcome may be very different and the surrogate mother may be forced to give up her child. Much discussion surrounding the Baby M case centered on the fact that the surrogate mother "broke" her contract with the biological father when she refused to give up her child. "She did not uphold her end of the agreement," "she broke her word," "she agreed to do this, so now she must keep her promise" were frequently heard statements. Careful review of surrogacy contracts shows that protection of surrogates' rights are not the major concern of such documents. Should society expect contracts to be enforced when those contracts are based on economic inequalities of the parties involved and lack provisions for the protection of individuals' rights? This question cannot be answered by lawyers alone; other members of society must express their personal values and help shape public opinion.

Maternal rights are the third unresolved question raised by the practice of commercial surrogacy. Does society have the right to force a mother to relinquish her child? This question becomes especially poignant when the basis for that removal is not maternal incompetency but a legally questionable document and a morally ambiguous practice. Surrogacy requires that surrogate mothers commit themselves to giving up the child before the fetus is even conceived. The women are denied the five-day cooling-off period after the child's birth that is required for women choosing to give their babies up for adoption. According to Barbara Katz Rothman, "surrogate parenting requires a woman to do what no other state allows to be forced on any mother in adoption: holding her to a promise made before the baby is born, to give up the baby forever" (1987). John Vetter, a psychiatrist and expert in the area of adoption, who has since 1965 counseled unwed mothers at an adoption agency, believes that a woman cannot know how she will feel about the baby until it is born. It is unrealistic to expect a surrogate mother to sign over her baby thirteen months before it is born, because "a vital piece of information is not possessed until the baby is born. There is no way to predict how strongly that bond will form, the nature of the bond at birth" (Port 1987g).

Implicit in arguments concerning maternal rights are issues that interweave biology and culture. If children could not only be conceived but also gestated outside of the human body, the questions would be less complex. Cultural roles, such as that of mother or father, are learned each generation anew by each member of the cultural group. These roles are not "instinctual" responses; individuals learn these roles as they learn the other cultural information required of them to become members of their group. All else being equal, individuals carry a similar potential for fulfilling particular social roles such as mother or father.

A child is not, however, carried to term outside of the human body. It grows in a symbiotic relationship with its mother; it responds to her voice and her heartbeat. During that period of gestation, her body creates a prenatal environment in which the fetus develops. That prenatal environment affects both the growth of the fetus and the woman carrying the fetus. Surrogacy requires a conceptual separation between social roles and biology, between parenting and pregnancy. What occurs during the pregnancy may influence how a woman feels about parenting. A surrogate mother enjoying her pregnancy may reconsider her decision to be excluded from parenting that child. The process of commercial surrogacy is dependent upon the surrogate mother's ability to terminate the emotional relationship simultaneous to the termination of the physical relationship, at the time of birth. If the surrogate mother is not able to end her emotional tie and turn the child over to the biological father and social mother, the court may be asked to decide who should be given claim to the child. The conflict occurs when one or another cannot effectively superimpose the conceptual distinction between social roles and biology.

Conclusion

Ethics are based on values and beliefs that reflect socially acceptable norms; as such, they tell us about ourselves. The controversy over surrogacy emphasizes an ethical dilemma in our society: Should we pursue the regulation and legitimacy of something technologically possible but which uses people as objects and commodities? When the rights of some individuals in the society potentially conflict with the rights of other individuals, whose rights should be protected? The concept of informed consent incorporates an ethical orientation toward the protection of a vulnerable population. This protection involves the right to know as much as possible about the implications of interventions. In the case of medical procedures, for instance, there is a responsibility to inform the subject population fully. In surrogacy, even though the medical consequences may be addressed, how can the sociocultural and psychological consequences be explained when they are, as yet, unknown? In surrogacy, can the unborn child, the surrogate mother and her family, the nurturing mother, and the social and biological father be protected from the future consequences of this intervention?

Singer (1985) suggested that we should try to regulate surrogacy even though he recognized that surrogate mothers may be unprepared to part with the child after it is born. Perhaps, like Great Britain, the United States should outlaw commercial surrogacy. That does not, however, solve the social and ethical issues inherent in the process. An article in a Hastings Center Report (Brahams 1987:1) questioned whether or not Great Britain moved too quickly in banning commercial surrogacy and suggested that a more limited regulation would have, perhaps, been more appropriate. It is clear that there is no consensus on how to resolve this controversy.

Commercial surrogacy is not going to go away just because it raises difficult social and ethical issues. It is now part of our contemporary reality, and we must undertake to protect those involved: the unborn children, women who choose to become surrogates and their families, and infertile couples. Based on the preceding analysis, two conditions emerge as means of protecting those involved in commercial surrogacy. The first condition is the provision of a grace period for the surrogate mother following birth, during which time she would be free to decide whether or not to surrender the child. The second condition, critical to the protection of surrogate mothers, is legislation establishing fair pay and working conditions for women choosing such a course.

In this chapter, I have focused on some of the legal and social issues raised by surrogacy. I have tried to highlight some of the ethical dilemmas emerging from commercial surrogacy and to provide an analysis of the social values involved. My aim in exploring the social and ethical issues surrounding surrogacy is to bring about discussion and argument, not to suggest specific solutions. Surrogacy merits thoughtful discussion and analysis; it changes the lives of all the people it touches. With five hundred surrogate-produced babies already born, and another hundred due this year, whether we agree or not with George Annas's opinion that "commercial surrogacy promotes the exploitation of women and infertile couples, and the dehumanization of babies" (1986:31), one thing is certain: We can no longer ignore it.

Acknowledgments

Many people have read this chapter as it evolved through its various stages; their comments helped move the writing forward, and the analysis was improved by incorporation of their ideas. Lois Randolph, Patricia Miller-Shaivitz, Kristen Leet, and Diane Rambeau each survived being my research assistant and contributed greatly to the completion of the book. My colleagues, Etta Binder Breit and Marilyn Myerson, each provided me with special insights from their own expertise. Doug Uzzell, neither student nor colleague, gave me the benefit of his critical evaluation and positive appraisal; Mary Floyd also provided a critical reading and concrete suggestions. The best of the chapter reflect their contributions.

Notes

1. This example is based on materials published by Susan Ince (1984) and Bernard Dickens (1984).

References

Annas, George J. 1979. Artificial Insemination: Beyond the Best Interests of the Donor. Hastings Center Report 9:14–15, 43.

______. 1984. Redefining Parenthood and Protecting Embryos: Why We Need New Laws. Hastings Center Report 14:50–52.

______. 1986a. At Law: The Baby Broker Boom. Hastings Center Report 16:30–31.

______. 1986b. At Law: Pregnant Women as Fetal Containers. Hastings Center Report 16(2):13–14.

Arditti, Rita, Renate Duelli Klein, and Shelley Minden, eds. 1984. Test-Tube Women: What Future for Motherhood? London: Pandora Press.

Bayles, Michael D. 1984. Reproductive Ethics. Englewood Cliffs, New Jersey: Prentice-Hall.

Blakeslee, Sandra. 1986. Some Caveats for Childless Couples. New York Times, April 27.

Brahams, Diana. 1987. The Hasty British Ban on Surrogacy. Hastings Center Report 17:16–19.

Brophy, Karen. 1982. A Surrogate Mother Contract to Bear a Child. Journal of Family Law 20:263–291.

Corea, Genoveffa. 1984. Egg Snatchers. Test-Tube Women: What Future for Motherhood? Rita Arditta, Renate Duelli Klein, and Shelley Minden, eds. London: Pandora Press.

______. 1985. The Mother Machine: Reproductive Technologies from Artificial Insemination to Artificial Wombs. New York: Harper and Row.

Crowe, Christine. 1985. "Women Want It": In-Vitro Fertilization and Women's Motivations for Participation. Women's Studies International Forum 8(6):547–552.

Dickens, Bernard. 1984. The Issue of Rights Associated with Alternatives in Conception. Paper presented at the Planned Parenthood Federation of America, New York, December 14–15.

Fee, Elizabeth, ed. 1983. Women and Health: The Politics of Sex in Medicine. Farmingdale, New York: Baywood Publishing Co.

Fletcher, John. 1981. The Fetus as Patient: Ethical Issues. Journal of the American Medical Association 246:772–773.

Fox, Robin. 1968. Kinship and Marriage in an Anthropological Perspective. New York: Penguin.

Frankena, William. 1975. Ethics. Englewood Cliffs, New Jersey: Prentice-Hall.

Freedman, Benjamin. 1975. A Moral Theory of Informed Consent. Hastings Center Report 5:32–39.

Gelman, David, and Daniel Shapiro. 1985. Infertility: Babies by Contract. Newsweek, November 4, pp. 74–75, 77.

Goetz, Sidney M. 1987. There Is Support for Surrogate Motherhood. St. Petersburg Times, February 18, p. 15A.

Goleman, Daniel. 1987. Motivation for Surrogate Moms Ranges from Guilt to Gratitude. St. Petersburg Times, January 25, pp. 1F, 3F.

Goodenough, Ward H. 1970. Description and Comparison in Cultural Anthropology. Chicago: Aldine.

Grobstein, Clifford. 1979. External Human Fertilization. Scientific American 240:57–67.

Holder, A. R. 1984. Surrogate Motherhood: Babies for Fun and Profit. Law, Medicine and Health Care 12:115–130.

Ince, Susan. 1984. Inside the Surrogate Industry. Test-Tube Women: What Future for Motherhood? Rita Arditta, Renate Duelli Klein, and Shelley Minden, eds., pp. 46–92. London: Pandora Press.

Juva, Mikko. 1985. Ethical and Moral Issues of Invitro Fertilization. Annals New York Academy of Sciences 442:585–587.

Keane, Noel P., and Dennis L. Breo. 1981. The Surrogate Mother. New York: Everest House.

Keane, Noel P., N. A. Pirslin, and C. S. Chadwick. 1983. Surrogate Motherhood: Past, Present and Future. Programme Clinical Biology Research 139:155–164.

King, Sherry. 1986. First Word. Omni (March), p. 6.

Lacayo, Richard. 1987. Whose Child Is This? Time, January 19, pp. 56–58.

Mackenzie, Thomas B., and Theodore C. Nagel. 1986. When a Pregnant Woman Endangers Her Fetus. Hastings Center Report 16(1):24–25.

Martin, Janice. 1987a. Pity the Judge for this Justice. St. Petersburg Times, January 7, p. 1B.

———. 1987b. Surrogates Are Really No Special Breed. St. Petersburg Times, February 16, p. 1B.

Milunsky, Aubrey A., and George A. Annas, eds. 1985. Genetics and the Law III. New York: Plenum Press.

Navarro, Vicente, ed. 1977. Health and Medical Care in the U.S.: A Critical Analysis. Farmingdale, New York: Baywood Publishing Co.

Parker, Philip J. 1982. Surrogate Motherhood: The Interaction of Litigation, Legislation, and Psychiatry. International Journal of Law and Psychiatry 5:341–354.

———. 1983. Motivation of Surrogate Mothers: Initial Findings. American Journal of Psychiatry 140:117–118.

Patterson, Suzanne M. 1982. Parenthood by Proxy: Legal Implications of Surrogate Birth. Iowa Law Review 67:385–399.

Poland, Marilyn. 1984. Influence of Reproductive Technology on Prenatal Responsibility. Unpublished manuscript.

Port, Robert. 1987a. The Battle over Baby M. St. Petersburg Times, January 4, pp. 1A, 12A.

———. 1987b. I Wanted to Have a Child of Our Own. St. Petersburg Times, January 7, pp. 1A, 4A.

———. 1987c. Whitehead: Stern Can Be Legal Father to Baby M. St. Petersburg Times, January 8, pp. 1A, 4A.

———. 1987d. Surrogate: I Never Meant to Hurt Them. St. Petersburg Times, January 9, pp. 1A, 8A.

———. 1987e. Judge Hearing Baby M Case Described as Caring, Careful. St. Petersburg Times, January 11, pp. 1A, 10A.

———. 1987f. Psychiatrist: Whitehead Should Never be Allowed to See Baby M. St. Petersburg Times, February 25, p. 3A.

———. 1987g. Problems with Adoptees Suggest Risk for Baby M, Expert Says. St. Petersburg Times, February 27, p. 3A.

_____. 1987h. Paternity Challenge, Fraud Allegation Further Complicate Case of Baby M. St. Petersburg Times, August 22, pp. 1B, 2B.

Powledge, Tabitha, and John Fletcher. 1979. Guidelines for the Ethical, Social and Legal Issues in Prenatal Diagnosis. New England Journal of Medicine 300:168–172.

Rothman, Barbara Katz. 1986. Commentary. Hastings Center Report 16(1):25.

_____. 1987. The Meaning of Choice in Reproductive Technology. Associated Press. St. Petersburg Times 2:23-1B.

Rowland, Robyn. 1985. A Child at Any Price?: An Overview of Issues in the Use of the New Reproductive Technologies, and the Threat to Women. Women's Studies International Forum 8:359–546.

Salvatore, Diane. 1986. Babies for Sale: White, Healthy and in Demand. Tampa Tribune-Times, July 20, pp. 1C, 7C.

Schroeder, Leila Obier. 1974. New Life: Person or Property? American Journal of Psychiatry 131:541–544.

Singer, Peter. 1985. The Ethics of the Reproduction Revolution. Annals New York Academy of Sciences 442:588–594.

Singer, Peter, and Deane Wells. 1983. Invitro Fertilization: The Major Issues. Journal of Medical Ethics 9:192–199.

Snitow, Ann. 1986. The Paradox of Birth Technology. Ms. 15 (December): 42, 44, 46, 76–77.

Snowden, Robert, and G. D. Mitchell. 1983. A Consideration of Artificial Insemination by Donor. London: George Allen and Unwin.

Snowden, Robert, G. D. Mitchell, and E. M. Snowden. 1984. Artificial Reproduction: A Social Investigation. Winchester, Massachusetts: Allen and Unwin.

Walters, Leroy. 1979. Human Invitro Fertilization: A Review of the Ethical Literature. Hastings Center Report 9:23–43.

The Warnock Committee Report. 1984. British Medical Journal 289:238–239.

Whiteford, Linda. n.d. Social Implications of Commercial Surrogacy. Ms. Paper presented at the 1985 annual meeting of the American Anthropological Association.

Winslade, W. J. 1981. Surrogate Mothers: Private Right or Public Wrong? Journal of Medical Ethics 7:153–154.

11

Surrogate Mothering in the Marketplace: Will Sales Law Act as Surrogate for Surrogacy Law?

Sandra Anderson Garcia

Through the use of rapidly advancing medical technology, at least six different means of achieving parenthood have been identified: artificial insemination by the husband (AIH), artificial insemination by a donor (AID), egg donation (ED), artificial embryonation (AE), embryo adoption (EA), and surrogate mothering (SM)[1] (Andrews 1981). Although all of these processes raise complex moral, ethical, legal, and psychosocial questions, it is the surrogate mothering arrangement that, to date, has posed some of the most perplexing legal issues involving parties' rights and responsibilities.

A surrogate is a person or thing that acts for, or takes the place of, another. The surrogate mothering transaction involves a contractual agreement between an infertile couple and a fertile woman who agrees to one of two arrangements: (1) "partial surrogacy," involving fertilization of the surrogate's egg through natural intercourse with, or artificial insemination by, the husband/donor or (2) "full surrogacy," in which the surrogate receives into her womb an embryo from an egg and a sperm, possibly the husband's and the infertile wife's, with which she has no genetic connection (Singer and Wells 1984). The fertile woman, acting as surrogate for the infertile wife, agrees that after she carries the fetus to term, she will allow the couple to adopt the baby.[2]

The surrogate, who may or may not be married, typically receives a substantial fee. The involvement of a third party (as opposed to germinal material, or eggs and sperm, only) throughout the entire transaction and thereafter and the exchange of large sums of money distinguish surrogacy from other collaborative reproductive methods and form the basis for unique legal problems. Moreover, the combination of these two factors makes the surrogate arrangement particularly vulnerable to commercialization and exploitation, as rights and responsibilities can be bargained for, bought and sold.

There is very little legal precedent to guide attorneys who bring together infertile couples and potential surrogates and attempt to draft contracts that will have legal effect. As of 1988, no state has enacted laws governing surrogate transactions[3] (Freed 1986). Thus, even as the parties to surrogate contracts and their lawyers look to the courts and to legislators for guidance, they must act in what is nearly a legal void. A rapidly evolving lexicon of surrogacy is emerging from that void—terms such as "baby brokers," "womb rental," "surrogacy stables," and "low-paid breeders"—and this terminology reflects the ad hoc behavior taking place. These terms also strongly suggest the commercialization of the process despite an almost universal rejection of a selling price attached to biological or social parenthood.

In the absence of laws that define terms and set standards for surrogacy arrangements, lawyers may be forced to turn to the law of sales for guidance. In this chapter, I examine some of the difficult social, legal, and ethical issues involved in this unique transaction.

The Commercialization of Surrogate Mothering

Article Two of the Uniform Commercial Code (UCC) is entitled "Sales" and covers modern mercantile sales transactions from the initial agreement through performance to discharge (Countryman, Kaufman, and Wiseman 1980). Article Two sets out the rights and responsibilities of people dealing in commercial transactions and reflects the rules and principles that have been developed by merchants since the Middle Ages. As a codification of the law of merchants, Article Two also contains the purposes, policies, and norms that merchants themselves have established as a basis for governing commercial behavior. Applying Article Two concepts to surrogacy may, at best, demonstrate the need to forestall and prevent the commercialization of the process and to enact and enforce surrogacy laws; at worst, it may encourage an excessive and inappropriate use of sales law principles, in which the terminology and standards of the marketplace are engrafted onto surrogacy.[4] In this worst-case scenario, unfertilized eggs might become known as "goods,"[5] the infertile couple depicted as the "buyer," the surrogate as the "seller,"[6] attorneys turned brokers who regularly arrange surrogacy deals for a fee would be "merchants,"[7] and a defective child might be considered "nonconforming goods"[8] (Countryman, Kaufman, and Wiseman, 1980).

The repugnance of such an eventuality is undeniable. However, as the terminology used to describe surrogacy and the behavior of parties involved in surrogate mothering transactions more closely approximate the behavior and terminology of the marketplace, fundamental changes in the process are likely to follow. The longer a legal void exists, the greater the likelihood of unchecked commercialization. Although some basic concepts and principles of sales law can, and should be, readily adapted to surrogacy, overall, the law of sales is too harsh and indifferent an instrument to optimally serve those involved in the surrogacy transaction.

Legal Issues in Surrogacy

There is no clear-cut line of demarcation between potential social-psychological problems and "purely" legal problems in surrogacy. For instance, questions related to whether a surrogate acted willingly after being fully informed of all the potential consequences of her participation involves her psychological state, on the one hand, and the legal issues of coercion, fraud, or misrepresentation, on the other. The surrogate's willingness to relinquish the baby is a similar issue involving motherhood, bonding, and attachment from the psychosocial perspective and breach of contract from the legal point of view. Although courts cannot be expected to redefine concepts as altruism, motivation, and maternal attachment, there are specific legal questions raised by the surrogate's involvement, in general, and her giving up the child she bore in exchange for money, in particular, that lawmakers must act upon.

Some questions that have reached the courts are: (1) whether the relationship between the husband/donor and the surrogate is adulterous[9] (Keane 1980); (2) whether the husband/donor has the legal right to claim paternity and to seek custody of the child born to a surrogate who is married[10] (Cohen 1985); and (3) whether surrogate motherhood contracts involving a fee paid at the time of adoption can be invalidated by statute or declared void as against public policy[11] (*Doe v. Kelley* 1981). In spite of this litigation, surrogacy is not currently governed by specific laws in any state and there is no consensus among those states that have introduced bills (Freed 1986).

Some of the still unanswered questions are (1) whether fetal rights exist; (2) whether the surrogate's obligations to ensure the safety and survival of the fetus she carried supersede her right to control her body; (3) whether adequate remedies exist in the event any party changes his/her mind and breaches the contract; and (4) whether surrogacy is a form of baby selling or merely a service. It is this last issue, sale versus service, that generates widespread debate and introduces semantic and behavioral maneuvering designed to bypass laws against baby selling.

Although there are laws against baby selling in every state, proponents of surrogacy claim that those laws do not apply to surrogate contracts. Such laws, they argue, were designed to protect poor, unwed mothers from unscrupulous persons who would pressure them to give up their children for money. Commenting on the inapplicability of black-market statutes to surrogacy, one observer writes:

> One difficulty in applying these statutes to surrogate motherhood contracts is that they relate only to the final phase—the adoption—of a contract which has other important provisions. The surrogate mother is not paid primarily for consenting to the adoption of the child. The services performed which justify substantial compensation consist rather of pregnancy and parturition, together with the risks and limitations which these experiences entail.
>
> . . . what the surrogate mother has to offer is what the sterile wife of the biological father unfortunately lacks: the biological capacity to reproduce (Keane 1980).

Although there is agreement among supporters of surrogacy that black-market statutes were not intended to apply to surrogacy, there is still disagreement among others. Some insist that the surrogate provides a service only, whereas others claim that labeling surrogate mothering a service is a distortion of reality because a sale does, in fact, occur. However, it is the surrogate's rights to her natural child that are sold, and not the child itself (Cohen 1985).

The issue of how much a surrogate should be paid to cover her expenses during pregnancy, the cost of delivery, and reasonable aftercare must be addressed by lawmakers. Further, they must rule on whether the termination of her parental rights upon receipt of her fee involves a sale of anything, including the child. Creating surrogacy laws and distinguishing them from laws governing adoption must also come out of the current legal void and not be left to those indulging in semantic, logical, and legal sleight-of-hand.

Meanwhile, where a transaction resembles a sale in terms of the terminology used, the behavior involved, and the expectations expressed and agreed upon, one will be hard-pressed to call it something other than a sale no matter how inappropriate such characterization might be. New and controversial medical techniques awaiting regulation and legal guidance have been forced to fit into preexisting law in the past. For example, during the first thirty years of artificial insemination by donor, the courts looked to existing laws regarding various aspects of family law in efforts to interpret and redefine parental rights and obligations (Billing 1985). With the increased and unchecked commercialization of surrogate mothering, parties and their lawyers may look to the law of sales for guidance to decide various legal issues.

The Sales Transaction

There are at least seven discrete stages through which the majority of sales transactions progress: preliminary negotiations (preagreement), agreement (contract formation), postagreement/preperformance (predelivery, for purposes of this analysis), shipment (delivery), receipt and inspection, payment, and postreceipt (Speidel, Summers, and White 1982). Surrogate mother transactions may be examined at each of these stages in terms of when particular events occur between the time of the initial contract negotiations and the ultimate handing over of the baby; at what points breakdowns in the transaction are most likely to occur; and whether fair and enforceable remedies exist for aggrieved parties.

Preliminary Negotiations (Preagreement)

Applying the law of sales to surrogacy at this stage would impose certain limitations on the parties' freedom to contract, such as the obligation of good faith in the performance and enforcement of every contract, conscionable dealings, and an absence of fraud, duress, and misrepresentation.

Good faith in the case of merchants means honesty in fact and the observance of reasonable commercial standards of fair dealing in the trade (Countryman, Kaufman, and Wiseman 1980). Good faith in surrogacy transactions seems essential, particularly in areas such as the surrogate's intention to prevent injury to, or destruction of, the fetus she will carry or her willingness to abide by the contract calling for the relinquishment of the baby for adoption. If a woman is shown to have been other than honest in these areas during preliminary negotiations, she has acted in bad faith and should be held liable for any negative consequences thereof. Proving the intent of the surrogate will present additional problems, of course.

Closely related to the concept of good-faith negotiations is the principle of unconscionability, involving prevention of oppression and unfair surprise. Negotiations are unconscionable in which there is no meaningful choice on the part of one of the parties, whereas the negotiated terms unreasonably favor the other party (Countryman, Kaufman, and Wiseman 1980).

Examples of unconscionable clauses would be ones stating that the surrogate alone can determine her behavior during the pregnancy without regard to the potential danger to the fetus. Illegal drug usage is an example of one such behavior. Other examples include clauses stating that upon receipt of a more lucratic offer, the surrogate can refuse insemination in spite of her binding contract, or that she can abort the fetus during the first trimester if she can enter into a more lucrative arrangement. Although it seems highly unlikely that any infertile couple would conclude negotiations with a potential surrogate who proposed such terms, the hypothetical clauses clearly illustrate the potential for bad faith and unconscionability at the negotiation stage.

One might assume that unconscionable clauses in surrogate contracts could most likely occur in situations in which the potential surrogate badly needs money, has inadequate or no legal advice, or in which an unscrupulous attorney/broker represents the infertile couple. However, with increasing commercialization and a growing number of middle-class, politically and legally sophisticated women negotiating surrogate contracts, the potential for inequitable terms in these contracts seems substantial, as the ability to negotiate for top dollar and maximum rights takes precedence over any altruistic motives the women might have.

Preliminary negotiations should be conducted in an atmosphere that will allow both the parental applicants and the potential surrogate to make competent, voluntary, and informed choices free from any attempts to defraud, coerce, or misrepresent. Thus, the Article Two mandates for good faith and fair dealing in sales transactions are principles that parties to surrogate mothering contracts should adopt, and that lawmakers should use in drafting all new reproductive laws.

The Agreement: Contract Formation

The surrogate mother contract is unique in ways that make this stage one of the least amenable to this Article Two sales analysis. Unlike the flexibility retail merchants enjoy, such as open terms over which they can later bargain and reliance on what the parties to the contract apparently intended, surrogate mother contracts should be as exact as possible. And unlike contracts between merchants, there is no partial performance in satisfaction of surrogate mothering contracts, particularly after conception has occurred. A merchant in oranges, for instance, may accept twelve crates even though he contracted for fourteen, agreeing to receive the two crates owed next season. The seller has partially performed. The surrogate must agree to carry the fetus until it can survive outside of her body, equaling full performance. Anything short of this would result in the death of the fetus.

Four basic terms of surrogate mothering transactions that cannot readily be left open for future bargaining are the prospective surrogate's willingness to: sell (or make available) her germinal material (egg) and reproductive capability; be inseminated by the husband/donor's sperm; agree not to intentionally abort the fetus when there are no serious threats to her own life; and agree to relinquish the baby after its birth. Equivocation in three of these areas present life and death questions for the child contracted for. Following conception, all attention must shift from any questions related to whether a valid and binding contract was formed, to matters concerning the best interest of all parties, especially those of the fetus who is the most vulnerable participant in the process.

These observations in no way suggest that surrogacy contracts can be 100 percent exact and binding. On the contrary, their complexity is such that accomplished drafters cannot include every term needed to spell out the rights and responsibilities of the parties and the remedies for breach of contract by any of them. Moreover, it is difficult to draft surrogate mothering contracts that do not pit the civil liberties of the potential surrogate against the interests of the infertile couple and the fetus. Such contracts have been described as basically "gentlemen's agreements" in that many of their provisions are unenforceable (Brophy 1981–1982).

This is particularly true for clauses such as those designed to: (1) control the surrogate's behavior during the pregnancy to ensure and monitor the well-being of the fetus (such as whether she can smoke, drink alcohol, take various medications, or if she will submit to fetal monitoring practices such as amniocentesis and ultrasound); (2) determine the method of delivery (whether she will agree to natural childbirth, a cesarean section if deemed necessary, spinal taps, and other interventions advised by medical personnel); (3) govern the actual handing over of the baby (the manner in which she will relinquish the child and her parental rights); (4) decide whether or not she will submit to specified tests to ensure that the baby was actually fathered by the man who contracted for it (the donor and husband of the infertile woman); and (5) settle how disputes will be resolved if the baby is defective and the surrogate's possible liability is at issue. Surrogate mothering contracts require far more specificity than do contracts for the sale of goods.

The contracts not only involve the constitutionally protected reproductive rights of the surrogate and the contracted for rights of the infertile couple; they also involve whatever rights the fetus might have. Therefore, such contracts demand clear and concise terms, a meeting of the minds of the parties, and adequate remedies for breach of contract. Whether or not such contracts can be drafted and enforced is a question that is not likely to be answered without new surrogacy laws and regulations. The law of sales cannot adequately provide the answer.

Postagreement: Preperformance

During the period between the signing of the contract and the time for the surrogate to relinquish the baby, parties to the contract might change their minds and either try to modify the contract or to repudiate it all together. Article Two provides that modifications must be signed writings that meet the test of good faith (Countryman, Kaufman, and Wiseman 1980). Sales law also outlines the actions a party to a contract can take when it appears that the other party will repudiate the contract or when he or she actually does so. Suspension of one's own obligations under a contract is one of the options parties have when breach of contract by the other party is expected or does occur[12] (Countryman, Kaufman, and Wiseman

1980). Between merchants, the primary concerns are how repudiation by the other party will affect their obligation to perform on the contract, the potential liability for their own repudiation, and which remedies are available to them if the other party wrongfully breaches the contract.

Attorneys drafting surrogacy contracts should apply the concept of good-faith modifications in order to discourage unscrupulous parties from attempting to extort unreasonable modifications of the original contract. However, the situation becomes more complex when demands for contract changes are coupled with threats to repudiate the contract. Such threats would leave a contracting couple in a surrogacy arrangement in an untenable position. For instance, a surrogate might raise her fee to an inordinately high amount midway through her pregnancy, without explanation or legitimate reason, and threaten to terminate the pregnancy if compliance with her demands were not forthcoming.

Options available to a buyer when a seller breaches a contract under the law of sales include the buyer's right to cancel the contract, to "cover" by buying substitute goods, to recover money damages for nondelivery, and, at times, to obtain specific performance whereby the seller must perform as contracted for. It is clear that only one of these remedies—specific performance—could satisfy a couple whose contracted-for fetus is being threatened with injury or death by a surrogate determined to disavow the contract through harmful acts or abortion. Specific performance, however, is the remedy least likely to be applied to surrogacy arrangements because it would introduce the repressive practice of forced gestation and delivery of the fetus.

Whether such compelled behavior will ever have legal effect or be enforceable must be left to speculation. However, the surrogate's ability to harm or destroy the fetus raises additional vital questions lawmakers must try to answer if fetal rights and the surrogate's right of personal privacy, including her decision to abort—as established in the landmark case *Roe v. Wade* (1973)—are ever to be reconciled.

Some such questions that demand answers are:

1. whether a surrogacy contract should be irrevocable once the surrogate has both accepted payment and allowed conception to occur
2. whether a surrogate contracts for a form of involuntary servitude for the nine-month gestation period during which others dictate which of her behaviors are acceptable and legal
3. whether a surrogate's decision to abort the fetus before or after its viability is legally any different from other abortions
4. whether a surrogate acquires the right to abort upon the unjustified breach of contract by the contracting couple, and if so, what and who determines the rights of the fetus under such circumstances

5. whether a surrogate's unjustified abortion (one not predicated on a danger to her health or survival) should be criminalized, resulting in her being charged with murder

Sales law cannot decide these and similar issues that might arise during the surrogate's pregnancy. The surrogate's desire to drastically modify or to breach her contract bears no resemblance to a merchant's tampering with his contract to buy oranges, refrigerators, or other goods. Applying Article Two remedies to surrogacy would bring about harsh, bizarre, and probably unconstitutional results—for example, compulsory cesarean sections. Notwithstanding the utility of applying the good-faith standard to contract modifications, Article Two provides few additional useful guidelines for this most crucial stage. Nonetheless, in view of the current void that exists in surrogacy law and regulation, Article Two concepts pertaining to buyers' and sellers' rights, responsibilities, and remedies might be inappropriately and destructively applied.

Delivery

The birth of the child occurs during this stage, and, because several things can go wrong, the concept of "risk of loss" may be borrowed from Article Two. There is an element of uncertainty in commercial sales transactions, and merchants who are aware that goods may be damaged, destroyed, or lost before or at the time of their delivery rely on provisions in Article Two that stipulate under what circumstances the risk of loss passes from the seller to the buyer. The legal significance of bearing the risk of loss and the possible application of the concept to surrogacy contracts may best be understood from the following description if "contracting couple" is substituted for "buyer," "surrogate mother" is inserted in place of "seller," "fee" replaces "price," and "baby" stands in for "goods":

> Ordinarily, risk of loss problems involve either an action for the price by the seller or an action by the buyer for restitution of the price or some part thereof already paid. If risk of loss was on the seller he/she cannot recover the price; or if the buyer has already paid the price the seller must return it. If the risk rested on the buyer he/she is liable for the price even though he/she may not have received the goods (Stockton 1983).

The concept of risk of loss can be assigned in the contract and, when applied to surrogate mothering contracts, involves three important primary factors that must be considered. They are: the possible death of the surrogate mother, the possible death of the natural father, or the birth of a defective baby.

Three clauses in a surrogacy contract drafted in Kentucky determine who will assume certain obligations in particular situations and may be considered as utilizing concepts similar to Article Two's risk of loss. The first clause concerns the possible birth of a defective child: "The natural father assumes the legal responsibility for any child who may possess congenital abnormalities and he has been previously advised of the risk of such abnormalities" (Brophy 1981–1982).

Concerning the possible death of the surrogate or the natural father,

> The natural father shall pay the cost of a term life insurance policy payable to a named beneficiary of the surrogate with a policy amount of $_____ and said policy shall remain in effect for six weeks subsequent to the birth of the child. In addition, the natural father shall make appropriate arrangements in his will for the support of the infant child should he die prior to the birth of said child and shall pay the cost of a term life insurance policy on his life payable in trust to said unborn child (Brophy 1981–1982).

The author of the clause comments that such provisions are necessary for the surrogate who has one or more children at home and that such an insurance policy is therefore mandatory. Another clause involving the risk of loss concept reads: "The surrogate and her husband understand and agree to assume all risks including the risk of death which are incident to conception, pregnancy, childbirth and postpartum complications" (Brophy 1981–1982).

There are additional issues involving the risk of loss concept as it relates to the payment of the surrogate's fee and to the acts of third parties invited into the surrogacy process. For instance, a surrogate needs to know whether, and under what circumstances, she assumes the risk of loss of her fee upon the death of the fetus during the pregnancy, or at birth, through no fault of her own. Parties to any surrogate mothering agreement would also want to determine whether, and under what circumstances, the risk of loss passes from the infertile couple to the surrogate upon the birth of a defective child. Finally, surrogate contracts should establish whether, and under what circumstances, the risk of loss passes to third parties invited into the process, such as genetic counselors, obstetricians, anesthetists, and others whose acts may result in the injury or death of the fetus.

The delivery of the child carries risks, some of which the surrogate may have to assume, whereas other risks may be contractually assigned to the natural father and his wife. The concept is one that surrogacy law should readily adopt with due consideration to the complex contingencies that exist among the surrogate, the couple, and the baby that may be born dead, alive and normal, or alive and defective.

Receipt and Inspection

> This, in commercial deals, is to seller and buyer alike, the crucial stage. Here the seller will typically lose any control he has over the goods. Here the buyer will, usually for the first time, inspect and decide whether to accept or reject what has been contracted for [Speidel, Summers, and White 1982, p. 638].

Article Two describes events that constitute acceptance of goods by the buyer, when such acceptance precludes subsequent rejection. It also details how the buyer may revoke acceptance of the goods in whole or in part (Countryman, Kaufman, and Wiseman 1980). The article provides that acceptance occurs when the buyer signifies to the seller, after a reasonable opportunity to inspect the goods, that the goods are conforming or, if they are nonconforming, that he will take them in spite of their nonconformity. Acceptance also occurs if the buyer fails to make an effective objection or does any act inconsistent with the seller's ownership, such as using or changing them (Countryman, Kaufman, and Wiseman 1980).

The issue of acceptance or rejection by the infertile couple in various situations raises several moral, ethical, and legal questions regarding the rights of the child and the responsibilities of the surrogate mother, the couple, and the state. Some of the questions that new surrogacy laws must answer are:

1. Under what, if any, circumstances may the contracting couple refuse to accept the baby because of a birth defect?
2. How long after the initial "inspection" does the couple have to revoke acceptance if a defect, or "nonconformity," is discovered?
3. Who will take the baby upon the death or divorce of the couple before the birth?
4. What liability rests with the surrogate mother for genetic defects she passed on if she knew, or should have known, of the possibility of such an eventuality—for example, Down's syndrome, Tay-Sachs disease, sickle-cell anemia?
5. What responsibility does the state bear to accept and care for babies who are rejected by parties to surrogate contracts?
6. Does the state have any choice in the matter?

At present, parties must rely on ad hoc solutions to these problems.

Although the inappropriateness of sales law is again demonstrated, once delivery has occurred, the concepts of conformity and nonconformity of that which was contracted for, acceptance of the "goods," and permissible revocation of the acceptance raise important issues when applied to surrogacy.

Attorneys must attempt to cover these matters in the contract. Enforcement without applicable laws in this area is a major problem, as I have tried to make clear.

Payment

Payment is an important issue during several stages of the surrogacy process. During preliminary negotiations and contract formation, lawyers must clearly spell out how much money is to be paid to the surrogate, for what it is being paid, and the time and manner payments will be made. Conciseness is necessary to ensure that surrogacy contracts are legal in the jurisdiction and to minimize the likelihood of conflict between the parties over fees. Without determined legal jurisdiction and fee arrangements, someone could repudiate the contract, with dire consequences.

Because of her ongoing sacrifices during the gestation and delivery, the surrogate mother is not likely to agree to only one lump-sum payment to be made when she gives up the baby. It is more probable that the surrogate will insist that her contract require an initial payment akin to a retainer, when insemination and/or conception occur. Another payment could be made after the birth, with the balance due when she relinquishes the baby for adoption. Whatever arrangements for payment are contemplated, the issue of laws against baby selling must be considered. Proponents of surrogacy transactions have attempted to defuse the issue by demonstrating the inapplicability of black-market statutes to surrogacy arrangements and by distinguishing between the receipt of a fee for reproductive capability and a fee for the handing over of a baby for adoption.

Payment is also an integral part of the concepts of "risk of loss" and the receipt, "inspection," and acceptance or rejection of a "nonconforming," or defective, baby. Article Two provides that a buyer is not excused from making payments if the contract requires payment before inspection for nonconformity of the goods unless the nonconformity is so obvious that it appears without inspection. However, such payment does not constitute an acceptance of the goods, impair the buyer's right to inspect, or deprive him or her of any remedies. These provisions shift some of the risk to the buyer because they require that he or she must pay first and litigate if defects are later discovered (Countryman, Kaufman, and Wiseman 1980). Surrogate contracts should contain these provisions in order to ensure that the surrogate is not deprived of her fee after the birth but before the receiving couple runs tests and performs other acts that would constitute an inspection of the baby.

With increased commercialization, the payment of fees for the eggs and reproductive capability of surrogates will raise issues such as how to prevent the exploitation of indigent women who are recruited by unscrupulous brokers. Further, the payment may raise the issue of the quality of the eggs, whereby eggs are categorized and priced according to race or ethnic group, age, or other factors. Such a practice could allow surrogates with "good," or grade-A, eggs to demand a higher fee than those owning eggs of lesser quality. Merchants who set the market value for surrogates may create a kind of genetic chauvinism or racism among potential buyers. This admitted speculation is presented to further stress the need to prevent commercialization and to call for regulation of surrogacy.

Payment of a substantial fee in surrogacy clearly raises complex questions even if those who assert that the process is a service and not a sale prevail. There is great potential for negative outcomes. Public policy, regulation, and new legislation are needed to answer perplexing questions about the role that the exchange of money will play in surrogate mothering transactions.

Postreceipt—Warranty Liability

In the terms of Article Two, the seventh and final stage in the surrogate transaction would involve warranty liability. Sales law provides different types of warranties: that goods will be merchantable when they are sold; that they are fit for the ordinary purpose for which such goods are used; and, in some cases, that the goods are fit for a particular purpose (Countryman, Kaufman, and Wiseman 1980).

Can a surrogate guarantee that her egg and her ability to carry and deliver a baby will produce a child that is acceptable to the natural father and his wife? Must she warrant that she does not have a physical condition—one that she is, or should be, aware of—that is likely to cause birth defects? If warranties can be issued in surrogacy, can they also be disclaimed or repudiated with words such as "as is" or "with faults" in the contract? It will be important for the contracting couple's attorney to be aware of words used to disclaim a warranty because such terms in ordinary usage of trade are understood to mean that the buyer will take the entire risk as to the quality of goods involved.

The harshness of sales law, where applied to surrogacy, is once more revealed. However, the notion that certain aspects of surrogacy should be guaranteed, such as an absence of deception when providing medical facts, seems reasonable. A remedy should exist for those who have been the victims of a breach of warranty. Whether any remedy would be enforceable is another issue.

Conclusion

By using the law of sales, I have pointed out in this chapter some ways in which the joy that an infertile couple experience when they sign a contract for a surrogate mothering procedure may turn into sorrow and loss. The couple may lose money invested and not get a child, or they may get a deformed child. They might be sued, if their attorney/broker does not draft a contract that adequately protects their interests. Without laws that regulate the surrogate mothering procedure, the possibility of pain and loss is also a reality for the surrogate mother and the child. A surrogate may be unjustifiably denied her fee, she may find it emotionally difficult to terminate her parental rights when she has donated her egg to help create the child, and she may suffer from guilt if the process results in a defective child. Fetuses and babies may become the focus of ongoing legal battles. The state may also lose by becoming the repository of deformed, unaccepted babies that are the products of breached and disclaimed surrogate mothering contracts.

That the law regularly lags behind the rapid advances of medical technology is a widely accepted premise. The degree to which the existence of this gap helps to shape, and sometimes corrupt, societal values and ethics, and how such a gap may even be detrimental to various members of society, should be studied and understood. Legislators and judges should use the knowledge of the negative effects of the gap between medicine's ability to alter human reproductive capabilities and the law's inability to regulate processes and mitigate the damage done as a basis for quick and decisive acts that will close the gap in cases concerning surrogacy.

Commercialization of surrogate mothering could put conceived but unborn fetuses and defective newborns at great risk. It is quite possible that they could become the innocent victims of entrepreneur attorney/brokers and enterprising surrogate mothers whose sole purpose would be to make money, and who would be able to operate with impunity because of the license of the law. One writer succinctly summarized this point: "Creation of children for money in the legal vacuum that exists today is a disservice to the children and to the legal system" (Holder 1985). With continued inaction by lawmakers, Article Two of the Uniform Commercial Code may partially fill the existing vacuum. And although certain concepts can be borrowed from the law of sales, Article Two concepts applied to the surrogate mothering process in response to its commercialization will, for the most part, be detrimental. Such application will introduce involved parties to the standards, customs, and language of merchants dealing with goods in the marketplace. The ultimate effects of such an eventuality on attitudes toward reproductive technology, in general, and surrogacy, in particular, is left to speculation.

Notes

1. In ED, an egg is taken from a woman other than the wife and implanted in the wife's uterus. Intercourse or artificial insemination by the husband results in

fertilization. AE involves the fertilization of a woman other than the wife by the husband's sperm. The embryo is flushed out of the woman and implanted in the wife, who bears the child. In EA, a fertile woman other than the wife is inseminated by a male donor other than the husband. The embryo is flushed out of the woman and implanted in the wife, who bears the child.

2. The wife of the donor, usually infertile or unable to carry a child to term for medical reasons, must adopt the baby because the law still views the woman who bears a child as its natural mother. However, with the use of "full surrogacy" there may be future litigation in cases in which the surrogate, though not genetically connected, may attempt to keep the baby, requiring new legal definitions of motherhood and the respective rights of the gestational (carrier) mothers, particularly as those rights relate to adoption and the termination of parental rights.

In at least twenty-eight states, the natural father may have to adopt the baby because those states have statutes that provide that offspring as a result of artificial insemination by donor is the legal child of the sperm recipient and the consenting husband, and the laws of fifteen of the twenty-eight states provide that the man who donates sperm to a woman who is not his wife is not the legal father of the child. For a listing of the states falling under both of the above, see "American Law and the New Reproductive Technologies," *Fertility and Sterility* (September 1986), p. 148, notes 28 and 29.

3. A number of state legislatures have introduced bills dealing with surrogate parenting aimed at regulating the arrangements. Also, the American Bar Association is formulating a "Model Alternative Reproduction Act" to regulate and validate reproductive technologies. Among the states that have been active in formulating surrogate legislation, called "alternative reproduction bills," are Alabama, Alaska, California, Connecticut, Hawaii, Illinois, Kansas, Maryland, Michigan, Minnesota, New Jersey, New York, Oregon, South Carolina, and Tennessee.

4. An analysis such as this requires an ongoing awareness of its purpose—to predict what might happen if the surrogate mothering transaction is thoroughly commercialized—so as not to become distracted by what might appear to be unacceptable concepts, such as labeling a woman's egg as "goods" that should carry a warranty or categorizing surrogate mothers and broker/lawyers as "merchants."

5. UCC §2-105 states that "(1) 'goods' includes the unborn young of animals, and (2) goods that are not both existing and identified are 'future goods'" (Countryman, Kaufman, and Wiseman 1980, p. 31).

6. UCC §2-103: "(a) 'Buyer' means a person who buys or contracts to buy goods, (d) 'Seller' is a person who sells or contracts to sell goods" (Countryman, Kaufman, and Wiseman 1980, p. 28).

7. UCC §2-104: "(1) 'Merchant' means a person who deals in goods of the kind or otherwise by his occupation holds himself out as having knowledge or skill peculiar to the practices or goods involved in the transaction" (Countryman, Kaufman, and Wiseman 1980, p. 29).

8. UCC §2-106: "(2) Goods or conduct including any part of a performance are 'conforming' or conform to the contract when they are in accordance with the obligations under the contract" (Countryman, Kaufman, and Wiseman 1980, p. 33).

9. Artificial insemination by a donor of a married woman not his wife has been considered to constitute adultery, which can render the surrogate mother contract unenforceable where adultery is a crime.

10. The natural father who seeks custody of the child born of a surrogate mother contract may face problems bringing suit in states that have adopted the Uniform Parentage Act, section 5 (b), which states that "the donor of semen provided to a licensed physician for use in artificial insemination of a married woman is treated in law as if he were not the natural father of the child." Further, a Kentucky court held that because the state law allowing a natural parent to terminate parental rights had been enacted specifically to facilitate adoption through licensed child-placement agencies and because its laws recognized a woman's husband as her children's father when they were married and had the capacity to conceive, it could not grant a petition from the biological father who sought custody of the child and the entry of his name on the child's birth certificate.

11. A Michican court upheld sections of the Michigan Adoption Code, including Mich. Comp. Laws Ann. §710.54 (West Supp. 1983–1984), which prohibited the father/donor and his wife from paying the egg donor, and surrogate mother, a $5,000 fee in conjunction with their use of the state's adoption procedures and which they tried to have declared unconstitutional. The court noted that "it is fundamental that parents may not barter or sell their children."

12. UCC §2-609 (1) provides: "A contract for sale imposes an obligation on each party that the other's expectation of receiving due performance will not be impaired. When reasonable grounds for insecurity arise with respect to the performance of either party the other may in writing demand adequate assurance of due performance and until he receives such assurance may if commercially reasonable suspend any performance for which he has not already received the agreed return" (Countryman, Kaufman, and Wiseman 1980, p. 120). UCC §2-610 provides that "when either party repudiates the contract with respect to performance not yet due, the loss of which will substantially impair the value of the contract to the other, the aggrieved party may for a commercially reasonable time await performance by the repudiating party, resort to any remedy for breach" available under the Code, or suspend his/her own performance (Countryman, Kaufman, and Wiseman 1980, p. 123). Suspension of performance of a pregnant surrogate would be a drastic remedy, bringing about the death of the fetus. Thus, suspension is not a realistic option for her.

References

Andrews, Lori. 1981. New Conceptions: A Consumer's Guide to the Newest Infertility Treatments. New York: Ballentine Books.

Billing, Ronda. 1985. High Tech Mothering: Legal Aspects of Biomedical Ethics. District Lawyer 9:56–63.

Brophy, Katie Marie. 1981–1982. A Surrogate Mother Contract to Bear a Child. Journal of Family Law 20:263–291.

Cohen, Barbara. 1985. Surrogate Mothers. Whose Baby Is It? American Journal of Law and Medicine 10:243–285.

Countryman, Vern, Andrew Kaufman, and Lipporah Wiseman. 1980. Article 2 Sales in the Uniform Commercial Code, 1972. Boston: Little Brown.

Doe v. Kelley. 1981. 307 N. W. 2d 438.

Freed, Doris J. 1986. As Surrogate Parenting Increases, States Must Resolve Legal Issues. National Law Journal 9, no. 15 (December 22, 1986), p. 28, col. 1.

Holder, Angela. 1985. Surrogate Motherhood: Babies for Fun and Profit. Case and Comment 90:3–9.

In Re Baby Girl. 1983. No. 83 AD Jefferson Cir. Ct. March 8.

Keane, Noel. 1980. Legal Problems of Surrogate Motherhood. Southern Illinois University Law Journal, no. 2 (1980), pp. 147–169.

Roe v. Wade. 1973. 410 U.S. 113.

Singer, Peter, and Deane Wells. 1984. The Reproduction Revolution. Oxford: Oxford University Press.

Speidel, Richard, Robert Summers, and James White. 1982. Commercial and Consumer Law. St. Paul: West Publishing Co.

Stockton, John. 1983. Sales in a Nutshell. St. Paul: West Publishing Co.

12

Surrogacy in the United Kingdom

Sybil Wolfram

In the parliamentary debates which led to the banning of commercial surrogacy in the United Kingdom in 1985 it was admitted that "commercial surrogacy has been successful in the United States" and that "we all admire many parts of the United States ethos" (*Hansard* 1985 [*6th series*], vol. 77, column 37)[1]: "[America] is a great innovator" (*Hansard* 1985, vol. 77, column 42). But it was also added that "we do not always try to emulate [the United States]" (*Hansard* 1985, vol. 77, column 37), "surrogate motherhood is one type of experiment that need never have arrived on our shores" (*Hansard* 1985, vol. 77, column 42–43), or, more viciously, "if this idea of commercial surrogacy has travelled across the Atlantic, it can return whence it came" (*Hansard* 1985, vol. 77, column 47).

These issues, and others relating to reproductive technology, were examined by a commission appointed by the government and chaired by Mary Warnock.[2] The commission's 1984 *Report of the Committee of Inquiry into Human Fertilisation and Embryology* (HMSO) became known as the *Warnock Report.* The word "surrogacy," which was given its present meaning in 1976 and came into common use in England in the early 1980s, is defined there as "the practice whereby one woman carries a child for another with the intention that the child should be handed over after birth" (HMSO 1984, sec. 8.1, p. 42). The *Warnock Report* does not tell us whether it is the woman who carries the child or the woman to whom it is handed after birth (or both) who is the or a "surrogate" mother. But in newspaper articles Mary Warnock applied the term to the carrying, or—as I shall dub her—incubating, woman (Warnock 1985b, p. 37), and in the 1985 act banning commercial surrogacy a " 'Surrogate mother' means a woman who carries a child in pursuance with an arrangement" (U.K., Parliament 1985, sec. 1).

In one way, "surrogacy" is not new and requires no technology. It has always been *possible* for a woman to gestate, or "incubate," a baby which she then hands to another woman. In particular, a man could have intercourse with a woman and agree that he and his wife should take the baby. After all, as Members of Parliament enjoyed pointing out, that is how Abraham's wife Sarah acquired a child in the Book of Genesis, and in the Highlands of Scotland sisters often enter into such arrangements (*Hansard* 1985, vol. 77, column 34). New technology, in this case artificial insemination and invitro fertilization, has merely, in the words of the *Warnock Report*, "eliminated the necessity of sexual intercourse in order to establish a surrogate pregnancy" (HMSO 1985, sec. 8.1, p. 42). Artificial insemination is, as it were, just a surrogate for intercourse. Invitro fertilization has added the new possibility that the ovum from which a child is born may come from one woman, who for ease of reference I shall call the "genetic" mother, and be inserted into another woman for purposes of incubation. The ovum could be fertilized by the genetic mother's husband, so that the child born is genetically not that of the incubating woman but of a married couple.

The Banning of Commercial Surrogacy

The *Warnock Report* was published in July 1984. Surrogacy hit the English headlines a few months later, with the birth of a baby known as "Baby Cotton." The baby was borne by a Mrs. Kim Cotton. She had been impregnated by artificial insemination from the sperm of a Mr. X, known only to be a foreigner. An American surrogacy agency had set up in England and arranged for Mrs. Cotton to be paid £6,500 (about $10,000) to incubate a baby and then hand it to Mr. and Mrs. X, who were to become its parents. The county in whose hospital Baby Cotton was born intervened to prevent the transfer, but Mr. and Mrs. X's lawyer made off with the baby, the courts supported their parentage, and the baby was taken to some destination abroad, later said to be the United States. The press carried pictures of Baby Cotton captioned "Born to Be Sold." The story, and arguments about surrogacy, became daily events on television. The *Warnock Report*, which had been commissioned by the Government to recommend what should be done about new reproductive technology and had come out about six months earlier, turned into a best-seller. Its chairman, Mary Warnock, a professional philosopher, became a Baroness and a household word overnight. Legislation against commercial surrogacy was produced by the government and passed by Parliament (U.K., Parliament 1985).

The *Warnock Report* was concerned with many issues besides surrogacy. Of the sixteen people making up the commission producing the report, six were qualified doctors and two more in medical administrative work. There were three lawyers, two social workers, one theologian, one philosopher (the chairman), and, curiously, one director of the Dartington North Devon

Trust. There was no social scientist, and the lawyers on the commission did not specialize in family law.[3]

Surrogacy was discussed mainly under the headings of "treatment" and "therapy" for the infertile and viewed as a way of alleviating some cases of infertility. The commission said that "the question of surrogacy presented us with some of the most difficult problems we encountered" (HMSO 1984, p. 46). Why was this so? The underlying reasons are considered below, but it is worth noting at once that the word "exploitation" occurs four times in the half page before the commission recommended: first, that all surrogacy agreements should be illegal contracts unenforceable in the courts and, secondly, that the creation/operation of surrogacy agencies, whether profitmaking or not—or, indeed, anyone knowingly assisting in the establishment of a surrogate pregnancy—should be liable to criminal proceedings. In February 1985—that is, after the birth of Baby Cotton—the Law Society went further and demanded that "it should become a criminal offence for a woman to offer, for reward of any kind, to bear a child for another, for a man or woman to offer such reward, and for a person to act as an agent or intermediary in such a transaction" (*The Times* 1985a).

Two bills, the Unborn Children Protection Bill and the Surrogacy Arrangements Bill, were introduced into Parliament almost at once. The former was a private bill, mainly dealing with experimentation with embryos, and failed.[4] The other, a government measure concentrating on commercial surrogacy, succeeded, and from July 1985 onwards anyone, except the incubating woman and natural father, involved in surrogacy arrangements for money was guilty of an offence, subject to imprisonment and fines.[5] By March 1986, there were attempts to close loopholes left by this act because "agencies in the United States are said to be recruiting British couples prepared to pay for a baby and women prepared to be surrogate mothers in exchange for payment" (*The Times* 1986a). *The Times* of May 15, 1986, reported avoidance in another form under the heading "Women Seek Loophole in Surrogate Ban": Two women were paid for surrogacy work in the name of research (*The Times* 1986b).[6]

Arguments About Surrogacy

Background on political argumentation and English law is necessary to disentangle the dispute about surrogacy in England. In political settings, arguments are often voiced not because they are necessarily believed but because they are effective in achieving or preventing a piece of legislation. These may be termed "political arguments."[7] English historical evidence shows that arguments of a certain *content* are peculiarly persuasive and occur on one side or the other whatever the particular subject under discussion, whether it is the divorce laws, laws about incest, allowing prohibited relatives to marry, or surrogacy.[8] We can call these "political *platforms*." The most common arguments, or platforms, in favor of a legislative change in England are that it is not really a change, or is a consequence of a change already made; that it opens to the poor what is open to the rich, or to women what is open to men; and that it brings England into line with Scotland (the two countries were unified only in 1707 and still have different private law). Conversely, effective arguments against change are that it is a change, will cause other changes, discriminates against the poor or women, or widens differences between England and Scotland.[9] The England-Scotland argument has not featured in the case of surrogacy: So far, legislation does not seem to include, nor explicitly to exclude, Scotland. I therefore loosely refer to them together as "England." There will generally also be arguments about social consequences specific to the issue in question, but although these seem important as a collective body—what you argue for should seem to have *good* and not bad social consequences—they tend to be indecisive, perhaps because they are usually fairly evenly matched. I shall reserve discussion of "moral" argument and a type of argument I distinguish as "category-based" until my analysis at the end of this chapter.

In English, as in U.S. law and belief, mother and father are related in the same way to the child—legally so, provided that they are married. Any child born to a married woman is presumed to be her husband's natural child, unless there is proof to the contrary. An unmarried woman who bears a child has the rights and duties normally shared by both parents, although the natural father, if known, has generally had some duties, mainly maintenance.

The adoption of children, originally chiefly war orphans, later mainly those of unmarried mothers, was legalized in 1926 after considerable discussion (U.K., Parliament 1926).[10] By 1954, when there was a commission on the subject, it was reported that adoption "had gained much greater importance in our social life than was expected in 1925" (HMSO 1954, p. 4, par. 19). It was generally performed by non-profit-making organizations. In 1958 it was made a criminal offence to make any payment, or give any rewards, for the adoption of children (U.K., Parliament 1958). In 1975 every local authority was obliged to set up an adoption service (U.K., Parliament 1975), although after the 1967 act legalizing abortion (U.K., Parliament 1967) there came to be a shortage of babies. (Welfare-style central or local government organization has been a very prominent feature of English life since the initiation of the so-called Welfare State after World War II—much more so than it is in the United States.) Adoptive relatives were assimilated by successive legislation to what are still known as "blood relatives." The old theory of heredity, prevailing until the beginning of this century, held that characteristics were transmitted equally from both parents "through the blood"—now, of course, by "through the genes" (see Wolfram 1987, ch. 1, sec. 1; ch. 7, sec. 1).

The arguments employed over surrogacy have to be seen against this background of existing law and what constitutes acceptable argument for and against change. The Warnock Committee itself recited under the heading of "Arguments Against Surrogacy" (HMSO 1984, secs. 8.10–12, pp. 44–45) that the weight of public opinion was against it. It would be an attack on the marital relation. To incubate another's child is "a greater, more intimate" service than that of semen donor (or ovum donor), that it is "inconsistent with human dignity" that a woman should use her uterus for financial profit, and that the strong bond between incubating mother and child would be breached.

Under "Arguments for Surrogacy" the Warnock Report (HMSO 1984, secs. 8.13–8.16, pp. 45–46) urged that surrogacy might afford a married couple their only chance of having a child genetically related to one or the other. Bearing a child for another can be seen "not as an undertaking that trivializes or commercializes pregnancy, but . . . as a deliberate act of generosity." Women have a right to use their bodies as they please, and, as to the bond of carrying mother and child, this has not been held to be an overriding objection to the mother placing the child for adoption.

These briefly recited arguments were developed in the press, in television discussions, and in Parliament. In favor of surrogacy, stress was laid on the plight of infertile women and the shortage of babies for adoption. Everyone, it was urged, has a right to a child; the infertile have the same rights as the fertile. Moreover, since artificial insemination is allowed so that a husband can procure a child legally his, it is unjust not to allow a wife to obtain a child which, as techniques improve, may even be genetically hers.

The arguments for banning surrogacy were amplified more. They are of two kinds. One affects all surrogacy. This is that legal motherhood should not be parted from bearing a child. English law has accommodated parting legal and biological fatherhood and does not normally regard either parent as "more closely" related to a child than the other. When this argument was expanded, as it was by Warnock, it had to be suggested that the concept of "relation" as applied to incubating mother and child carried the overtone of "feeling for" and "physical proximity to" (Warnock 1985, p. 3). This argument would, of course, have equal force against noncommercial as well as commercial surrogacy, but it is sometimes conjoined with the idea that if "surrogacy" is performed "for love"—as, for example, between sisters or friends—that alters the situation.

The second set of arguments against surrogacy is specifically against commercial surrogacy and hinges on money. It is noticeable in the *Warnock Report* that the words "donor" and "donate" and "therapy" are in constant use. A man may "donate" sperms or a woman ova for use in "therapy" that will enable a woman to bear a child if her husband is infertile or if she herself cannot produce ova. The uncomfortable fact that semen used in artificial insemination was paid for was explained as being just a nominal thank you to the man for his trouble, but it was recommended that payments should be phased out and that a woman should receive no compensation for ova extracted from her to insert into another woman or for other purposes such as research (HMSO 1984, sec. 27, pp. 27–28). However, while sperm and ovum donation seems to be within the pale, provided no money passes, the donation or loan of uteruses or handing over of babies is beyond it.

Opponents of surrogacy generally speak of it as a "buying" or "selling" of babies, or, if not that, then as a "renting out" of uteruses (e.g., *Hansard* 1985, vol. 77, columns 37 and 42). It is unacceptable, one Member of Parliament said, "to offer one's womb for reward" (*The Times* 1985b). "There is something wrong," another said, about the "purchase of babies" (*Hansard* 1985, vol. 77, column 34). There is "moral repugnance," Warnock said in a radio talk, "to the thought of a woman deliberately becoming pregnant for money" (Warnock 1985, p. 3).

We have to note that the sums of money involved are relatively large. Thus it was possible to argue that not only was commercial surrogacy as such morally repugnant but also it would open means of procuring babies to the rich which would not be open to the poor. And terrible abuses might follow. For instance, a woman who does not want the bother of pregnancy or dislikes the idea of giving birth to a child might, if sufficiently wealthy, tempt a poor woman to do this for her, thus creating a slave class of women. And what if when the baby is born, it is deformed and undesirable? When this style of argument is used against commercial surrogacy, it is sometimes linked with the suggestion that some degree of noncommercial surrogacy should be introduced—not merely allowing surrogacy provided no money passes, but setting up controlled noncommercial organizations, like those arranging adoption, to supervise it and see that it is performed in accordance with satisfactory rules.

Surrogacy and Money

The U.S. commercial agency which arranged for Mrs. Cotton to act as surrogate mother for Baby Cotton argued that no one got rich on their arrangements. The surrogate mother just got "compensation" for her trouble, and the agency, which put in a lot of work, got a normal, by no means exorbitant, fee. I have heard it suggested in the United States that some of the objections had arisen from the fact that the agency was American. What was said was rather that the United States was a much more commercial society than England, where anything, even babies, could be bought and sold. Certainly there seems to be a difference between the two societies in this area.[11]

The objection in England seems to split into two. One is that as surrogacy is expensive to would-be parents, the rich gain an advantage over the poor, an effective political argument. However, as the rich are not prevented from buying yachts or other items not available to the poor, the use of this argument is either purely factitious—that is, simply designed to get and keep commercial surrogacy banned—or it is founded on some further argument, such as that the rich should not have privileges with respect to acquiring babies. It was possibly to counter this that friends of commercial surrogacy urged that the fertile should not have advantages over the infertile, nor infertile men over infertile women, and/or that it is unfair that those with relations or friends willing to bear babies for them should be in a favored position with respect to getting babies.

The other objection to money passing for babies is different. This is that babies are not a commodity for which money should pass. "Moral repugnance" to money passing for babies is expressed in nonpolitical settings—where voicing it does not help alter anything—and it does seem that there is genuinely thought to be something wrong with it. I question the use of the word "moral" in this setting. What we have here I think is what I call a "category-based argument," seldom made quite explicit. In England, more so than in the United States, it is considered that only things or animals may change hands for money. To allow money to pass for people is to reduce them to "things" or animals. The treatment of sperms or early embryos as things or on a par with animals is just tolerable. But this does not extend to babies, who are thought of as already people.

It is not really the case that if money passes in a surrogacy situation a baby becomes a thing. After all, in every other respect babies are subject to the laws governing people and not to the laws governing things or animals. And it seems wholly improbable that admitting commercial surrogacy would lead to slavery or anything resembling it. The objection seems due simply to the marking off of the human category as a category of "object" which *inter alia* cannot be bought and sold (just as it cannot, must not, for example, be eaten).

Advocates of surrogacy do not generally try to convince people that there is not really anything wrong with babies changing hands for money. Rather, they suggest that this is not what is going on. Money is used to buy and sell services as well as things or may be simply given, as a present. But suggesting that the money that passes in surrogacy situations falls into these categories does not help advocates of surrogacy. First, sexual style favors are not properly bought and sold. Secondly, the recipient of money is normally the inferior of the giver of the money. This is particularly noticeable over sexual services, where the recipient is labeled a prostitute.

But receiving money is an inferior thing in other situations too. It would be insulting to hand a fiver or 50p piece to a superior—say, one's tutor or boss—in a way that it is not to tip a waiter or a taxi driver or a child. And it may be to avoid implications of superiority and inferiority that the *Warnock Report* used the term "donate" rather than "give" with respect to the giving and receiving of sperms and ova of which it approved. But to give or offer money, or expensive presents, is a much more striking way of implying superiority. This is again obviously no fact but a local indigenous association. It leaves advocates of surrogacy in a poor position. They can argue that no sex is involved, since the babies are otherwise created, so that money is not passing for strictly sexual services. They can say that wombs are not being hired out. The incubating woman is just being given some pin money as a present or a thank you, but that makes her some kind of inferior.

I suggest that it is these associations of commerce that makes surrogacy with money passing so unpalatable to the English soul. However the money passes, whether for the baby, for the service, or as a present, it still does not seem in English eyes an acceptable passage of money. So commercial surrogacy seems "morally repugnant" enough to be outlawed.

Surrogacy not for money still has problems. One is perhaps minor. This is that to incubate a baby for nine months and to give birth to it for someone else is a very substantial service. However, sisters or friends may do much for one another, especially in times of trouble. To bear a baby for a sister or friend may be brought under this umbrella. The other problem of noncommercial surrogacy arises from the kinship angle. It raises, as Warnock had it, "the stark issue of who is the mother" (HMSO 1984, sec. 6.8, p. 37), an issue which has not previously arisen but is almost bound to raise its head because of the new techniques which allow a woman to bear a child from an ovum not her own—or, as it is sometimes phrased, which is not genetically hers. This means that it is no longer obvious that the woman bearing the child is the biological mother, nor therefore that she should always in future be the legal mother. In my opinion, noncommercial surrogacy will be resisted, but perhaps because of the analogy with fatherhood (where a man may be father to a child to which he is not biologically related by being married to the mother of a child), it will ultimately be accepted. It seems probable that it will at least initially be veiled under such terms as "therapy for infertility" and thus made to pass as belonging to an acceptable category: viz., treatment for illness or incapacity. But its acceptance does not seem at all certain. One reason for this doubt reflects current problems in society about the equality of men and women.

The issue may well shift through further technical advances. The fact that some women are unable to incubate or bear children might be overcome by technology enabling them to do so. Alternatively, technical advances of another kind, such as the improvement of machine incubators, so that the uterus is reduced to the "natural" as opposed to "artificial" incubator of a child, may solve the dilemma. This latter solution would, in theory, enable women who preferred it to have children genetically their own without having to incubate and bear them. An ovum could be extracted, mixed with the husband's sperm, and the resulting embryo hatched by artificial incubation, as chicks have been for many years. Repairing reproductive organs so that women now unable to bear children would be able to do so would certainly be approved. But incubation by artificial means, even if perfected, would almost certainly raise moral hackles in the United Kingdom. One might ask whether this is only because it appears an unnatural novelty or whether there is more to it, and if there is more to it, what are these underlying social values?

Acknowledgments

Earlier versions of this chapter were presented to the Intercollegiate Seminar, University of London, Goldsmith's College, Department of Anthropology, in November 1985 and at the Annual Meeting of the Society for Applied Anthropology in Reno, Nevada, in March 1986, and I am grateful for discussion and helpful criticisms. I should also like to thank the Board of Literae Humaniores of the University of Oxford for a grant assisting me to attend the meeting at Reno.

Notes

1. See *Hansard* 1985 (6th series), vol. 77, col. 37. In England, legislation requires three "readings" of a bill, with a majority in favor in each in both Houses of Parliament (the House of Commons and the House of Lords), followed by "Royal Assent." The critical reading with the most extensive debates is normally the Second Reading in the House of Commons, from which I quote here. Parliamentary Debates are recorded verbatim in what is known and cited here as "Hansard."

2. HMSO 1984. Such commissions, technically by "Command of His/Her Majesty" and therefore called "Royal Commissions," are appointed by the government to find facts and make recommendations when certain styles of proposed legislation are under consideration.

3. See *Warnock Report* 1984, pp. ii–iii for the list of members.

4. See, e.g., *The Times*, June 8, 1985, p. 4, cols f–h. Some parliamentary time is allowed for bills introduced by individual Members of Parliament. These are known as "private bills," and in general stand less chance of success than bills introduced by the government—i.e., under the auspices of the political party with a majority in the House of Commons, which includes the Prime Minister and his/her ministers, the more important of whom form the Cabinet. A defeat for a government bill can spell the end of the government's power and result in a general

election. In the case of the Surrogacy Arrangements Bill a "free" vote was allowed—i.e., not along party political lines.

5. The Surrogacy Arrangements Act 1985 set the penalty, fairly modestly, at three months' imprisonment or £2,000/(about $3,000) fines.

6. The matter was explained clearly in *The Times* May 15th 1986. "Two would-be surrogate mothers pregnant through artificial insemination intend to hand over the babies to their natural fathers, the husbands of infertile women, in spite of a ban on the practice. The recipients are paying £13,000 each [about $20,000], and the surrogate mothers will receive between £5,000 and £7,000 for keeping diaries on their experiences.

Miss Lorrien Finely, who heads Reproductive Freedom International which organized the arrangement, claims she has not broken the law because the surrogate mothers are being paid for taking part in research during their pregnancies rather than for selling their babies."

See the later section on "Surrogacy and Money" on acceptable payment of money. Payment for research is perfectly acceptable. It might, however, in this case well be thought that the object of the exercise is to make surrogacy a *fait accompli*, so that it can be argued that it would no longer be a change to allow it.

7. It is arguable that arguments we voice or more generally anything we *say*, as opposed to what we may believe, is always for some purpose, to secure some end or other (see Wolfram 1985a: 71–84). What I term "political arguments" are distinguished by the fact that the object of voicing the arguments in question is not just, for example, to pass the time of day or clear one's head or advance learning but to effect or prevent a particular practical change.

8. This theme is discussed in relation to English kinship, or family law, during the nineteenth and twentieth centuries in Wolfram 1987. Chapter 11 is concerned with surrogacy and related issues. For examples, see also Wolfram 1983: 308–316; Wolfram 1985: 155–186.

9. The platforms mentioned here relate to legislation. Other English settings call forth different ones. For example, in an Oxford college a similar role is performed by: "We must think of the undergraduates."

10. The Adoption of Children Act 1926. It was preceded by two royal commissions: *Report of the Committee on Child Adoption* 1921 and *First Report of the Child Adoption Committee* 1925.

11. "In America You Can Buy Anything—Even Babies." TV, Channel 4, "Diverse Reports: Cash on Delivery," January 30, 1985.

References

Hansard. 1985. Parliamentary Debates, 6th Series, 1984–1985, vol. 77, House of Commons.

HMSO. 1921. Report of the Committee on Child Adoption. Cmd 1254.

———. 1925. First Report of the Child Adoption Committee. Cmd 2401 London.

———. 1954. Report on the Adoption of Children. Cmd 9248 London.

———. 1984. Report of the Committee of Inquiry into Human Fertilisation and Embryology. Cmd 9314 London (Chairman M. Warnock).

The Times (London). 1985a. February 4, 1985:3 b-c.
______. 1985b. April 16, 1985:3 a-b.
______. 1986a. March 10, 1986:3 a-b.
______. 1986b. May 15, 1986:3a.
U.K., Parliament. 1926. The Adoption of Children Act, 16 & 17 Geo 5 c. 29.
______. 1958. Adoption Act, 7 & 8 Eliz II c. 5.
______. 1967. Abortion Act, c. 87.
______. 1975. Children Act, c. 72.
______. 1985. Surrogacy Arrangements Act: An Act to regulate certain activities in connection with arrangements made with a view to women carrying children as surrogate mothers [16th July 1985], Eliz II c. 49.
Warnock, Mary. 1985a. The Surrogacy Scandal: Legal Surrogacy—Not for Love or Money? The Listener January 24:3.
______. 1985b. When Mother Makes Three. Sunday Times July 14:37.
Wolfram, Sybil. 1983. Eugenics and the Punishment of Incest Act 1908. Criminal Law Review (1983):308–316.
______. 1985a. "Facts and Theories: Saying and Believing." In Reason and Morality (ASA Monograph 24). J. Overing, ed. Pp. 71–84. London: Tavistock.
______. 1985b. "Divorce in England 1700–1857." Oxford Journal of Legal Studies 5(2):155–186.
______. 1987. "In-laws and Outlaws. Kinship and Marriage in England. London and Sydney: Croom Helm; New York: St. Martin's Press.

13

The Baby M Case: A Class Struggle over Undefined Rights, Unenforceable Responsibilities, and Inadequate Remedies

Sandra Anderson Garcia

This chapter is designed to raise serious legal, psychosocial, and ethical issues about surrogate mothering arrangements that were involved in the events that led up to the Baby M trial. I will argue that such transactions are so complex, and that there are so many competing interests, that it is virtually impossible to define clearly and justly the rights and responsibilities of all involved parties or to find adequate legal or equitable remedies for those wronged when transactions break down. Extreme class differences between surrogate mothers and infertile couples are identified as further complicating the process by increasing the opportunity for exploitation and inequity.

The word "parties," as used in this chapter, includes not only the surrogate mother and the infertile couple with whom she makes an agreement but also the fetus, the husband and children of a married surrogate, the wife and children of the married sperm donor, the child born of the arrangement, sets of grandparents, and the state within whose jurisdiction the process occurs. Such a broad identification of parties is intended to show that at this early stage in the development of surrogacy, we have failed to take four crucial steps: (1) to adequately identify the parties who justifiably have an interest in surrogacy transactions and in the children born of those transactions; (2) to identify, study, and argue for the rights and responsibilities of every party of interest; (3) to examine the remedies each party might have; and (4) to consider implications of great socioeconomic difference between the contracting parties.

I will allude to testimony from the Baby M trial primarily to illustrate a point that I made in Chapter 11 concerning the possible commercialization of the surrogacy process and to introduce issues that may be anticipated in future surrogacy trials.

The Baby M case allows us to move from pure speculation and prediction to an empirical perspective. It is readily acknowledged that one case in no way presents adequate data upon which definitive statements can be based. However, adequate data does not exist. To my knowledge, to date only a few surrogate mothering transactions out of over five hundred carried out have been litigated.[1] These few cases are precedent-setting, and it is hoped that this overview of the Baby M case will promote analysis and constructive discussion of the many complex issues involved. I would also strongly urge the speedy passage of legislation in all states that will either outlaw or closely regulate this extremely complicated human interaction that we call surrogate mothering.

Background

A successful surrogate mothering arrangement between a fertile woman and an infertile couple may proceed smoothly from initial contact of the parties to the ultimate adoption of the baby born of the arrangement by the infertile woman. Some transactions are between family members or friends, and in these cases, there are usually no lawyers, no formal contracts, and few problems.[2] The vast majority of transactions between strangers brought together through infertility centers also usually conclude successfully. Depending on the state's adoption laws, the successful process can take less than one year.

Where there are major problems, however, such as the breach of the surrogacy contract by either the couple or the surrogate, the process may involve legal maneuvering, bargaining, threatening, and expensive litigation over custody of the baby. This scenario occurred in the Baby M case. In that case, three years after a potential surrogate mother, Mary Beth Whitehead, answered an ad with the intention of becoming a surrogate mother, custody of the baby born of the arrangement that she entered into had not yet been definitively decided.

The chronology that follows presents significant events that preceded the lengthy court battle in the case and sets the stage for the seven-stage analysis that follows it. However, it is first worth noting that the use of certain terminology was widely discussed during the case. Because a surrogate acts for, or takes the place of another, it was argued that Mrs. Whitehead has been incorrectly called a surrogate mother when, in fact, she is the biological mother and a surrogate carrier/gestator. She acted for, or took the place of, Mrs. Stern.

Word usage has become an issue primarily because of questions related to whether the egg donor, the gestator (or woman who carries the baby), or the social mother (a third woman who might ultimately adopt the baby) is the legal mother, notwithstanding contract terms. These questions of definition are unanswered as of this writing. They will persist because of the ambiguous legal definition of motherhood, a critical element required for delineating women's rights and responsibilities.

Chronology

March 1984

Mary Beth Whitehead, a twenty-seven-year-old married mother of two answers an advertisement in a New Jersey paper and goes to the Infertility Center of New York, a for-profit agency, where she applies to become a surrogate mother. The center's director and founder is Noel Keane, a lawyer. He has closed at least 150 surrogacy "deals" and has been called the "father" of commercial surrogate motherhood.

December 3, 1984

William Stern, a thirty-eight-year-old biochemist married to a pediatrician, signs an agreement with the Infertility Center of New York to seek a surrogate mother to bear his child.

January 1985

The Sterns and the Whiteheads meet at the center.

February 6, 1985

The Whiteheads and Dr. Stern sign a twenty-page contract providing that Mary Beth Whitehead will be artificially inseminated with Dr. Stern's sperm and that at birth she will give the baby to him and his wife so that they may adopt the child. She will then receive a $10,000 fee. The contract is signed at the center.

July 1985

After six months of artificial insemination, Mary Beth Whitehead conceives.

July 1985–March 1986

During the period of gestation, the Whitehead and Stern families become friends. The Sterns bring the Whiteheads gifts after a trip to Switzerland and occasionally entertain the Whitehead's two children. Mrs. Whitehead, while pregnant, gains 50 pounds and has phlebitis and other complications.

March 27, 1986

Mary Beth Whitehead gives birth to a 9-pound, 2-ounce baby girl at Monmouth Medical Center in Long Branch, New Jersey. Mrs. Whitehead names the baby Sara Elizabeth; the Sterns name her Melissa Elizabeth. Court papers will later call the child Baby M.

March 30, 1986

Mrs. Whitehead gives the baby to the Sterns.

March 31, 1986

Mrs. Whitehead drives to the Sterns' home and in great distress begs to have the baby returned to her. Fearing that Mrs. Whitehead might harm the baby and herself, the Sterns give Mrs. Whitehead the baby. Mrs. Whitehead nurses the baby.

April 12, 1986

When the Sterns visit Mrs. Whitehead, she says that she cannot give up the baby.

May 5, 1986

Judge Harvey Sorkow of Bergen County, New Jersey, grants the Sterns temporary custody of Baby M. According to police accounts, Mrs. Whitehead hands the baby out of a bathroom window to her waiting husband and that night they flee to her parents' home in Pasco County, Florida.

July 15, 1986

Mrs. Whitehead phones Dr. Stern and threatens to kill herself and Baby M, stating, "I gave her life. I can take her life away." Dr. Stern tapes the call.

July 16, 1986

Mrs. Whitehead phones Dr. Stern and threatens to accuse him of sexually molesting her ten-year-old daughter if she is not allowed to keep the infant.

July 31, 1986

The Sterns hire private detectives, who locate Baby M in Florida. While Mrs. Whitehead is hospitalized with a kidney infection, sheriff's deputies take Baby M and return her to the Sterns.

August 1986

After Mrs. Whitehead returns to New Jersey with her family, Judge Sorkow orders paternity tests after Whitehead alleges that she had sex with her husband during the same period of time that she was being artificially inseminated and that the baby may be his. The Sterns' lawyers submit Florida Court documents declaring Dr. Stern the father of Baby M based on affidavits signed by the Whiteheads and Dr. Stern. The Whiteheads sue to strike down Stern's paternity judgment, alleging fraud.

September 4, 1986

A Florida judge sets aside the Stern paternity judgment on a technicality.

September 10, 1986

Judge Sorkow grants Mrs. Whitehead visitation rights with Baby M for an hour twice a week at a supervised foster home. Her attorneys ask for proof that Mrs. Stern is infertile, claiming that if she is not, then the contract between Dr. Stern and Mrs. Whitehead is invalid. Mrs. Whitehead's visitation rights are changed to two hours once a week.

September 25, 1986

Paternity tests show that Dr. Stern is almost certainly the father of Baby M and that Mr. Whitehead cannot be the father.

October 7, 1986

Judge Sorkow issues a gag order forbidding parties in the case to discuss the reasons for Elizabeth Stern's infertility or releasing transcripts of taped phone conversations in which Mrs. Whitehead allegedly threatened to kill herself and Baby M.

October 14, 1986

Mrs. Whitehead sues the Infertility Center of New York, its executive director, the doctor who performed her insemination, and two Florida lawyers involved in the case. The suit asks that her contract with the Sterns be invalidated and that punitive and compensatory damages be awarded for the emotional distress she suffered. Whitehead charged in her suit that personnel at the Infertility Center of New York failed to counsel her adequately about surrogate parenting and neglected to advise her to have an attorney look over the contract. Fraud was also alleged in connection with a paternity proceeding in Florida that was seeking to have Dr. Stern named as the father of Baby M.

November 13, 1987

After several delays, a trial determining the legality of the contract between Mrs. Whitehead and Dr. Stern is scheduled to begin.

January 5, 1987

The custody part of the trial begins. Judge Sorkow states that he will withhold a decision on the validity of the contract until he hears and evaluates the testimony in the custody phase of the trial.

March 12, 1987

Closing arguments are presented.

Prenegotiation

There are at least five areas to consider during the period before the potential surrogate mother and the infertile couple initiated the surrogacy process: (1) the adoption statutes and anti-baby-selling statutes in the jurisdiction in which a disputed transaction would be litigated; (2) laws that determine the legal parents of the child born of a surrogate arrangement; (3) the expertise, ethics, and credibility of the personnel in an infertility agency or of an individual broker the parties plan to use to arrange the transaction; (4) the motivation of the contracting parties; and (5) the mental and physical health of the surrogate and the couple.

Preliminary Negotiations

Key players at this stage are the person who brings the parties together, the lawyers for both parties, and psychological consultants and social workers who act as advisors.

During preliminary negotiations, facts are presented by the broker, ground rules are set, and parties' expectations are established. Parties are expected to negotiate in good faith and without duress, fraud, misrepresentation, oppression, or unfair surprise.[3] Some areas covered are: psychological and physical tests the surrogate mother will be asked to undergo, the reasons why the parties are entering into the arrangement, parties' willingness to divulge relevant medical histories, and the surrogate's acknowledged understanding that she will relinquish the baby and terminate her parental rights. Nondisclosure or deception during preliminary negotiations can create future problems.

An example of how lack of candor during preliminary negotiations in the Baby M case became an issue during the litigation is the allegation by Mary Beth Whitehead's attorney that the contract should be unenforceable because of Mrs. Stern's representation that she is infertile when, in fact, she is not. This issue seems to support the point that there is considerable need for informed and conscientious negotiations before a contract is drafted. A more astute negotiator might have inquired at length into Mrs. Stern's alleged infertility and into its significance to the transaction.

Contract Formation

The drafting of the surrogate mothering contract is critical, particularly at this stage in the legal development of the process. If such contracts are held to be enforceable and not against public policy, it will be imperative for attorneys to try to include the maximum number of terms covering every significant and forseeable eventuality. The terms must be conscionable.

As the number of attorneys specializing in surrogacy laws increases, their services will be in demand not only at the stage of contract formation but also during litigation over enforcement of contract terms. The party able to employ these specialists will have a decided advantage during this period at which rights and responsibilities are not yet clearly defined. A case in point is a clause in the Whitehead/Stern contract that stipulates that if the child is stillborn, Mrs. Whitehead will receive only $1,000. Whether this clause is "fair" and conscionable or blatantly unjust is debatable. When contracts are drafted by attorneys with vastly different levels of expertise, such terms can become a part of a contract without a challenge.

Enforcement of contract terms might be problematic no matter what a party's socioeconomic status, because the surrogate mother can claim a right to control her body and act in ways that breach contract terms and threaten the safety and survival of the fetus at the same time. If she is willing to take the consequences of such acts, it may be pointless to argue about her responsibilities to the fetus or to search for satisfactory remedies.

Although this was not an issue in the Baby M case, it would appear to be important to define the responsibilities to the fetuses of all surrogate mothers, and to investigate the most effective ways to impress upon them the importance of those responsibilities, duties, and obligations. It seems a logical consideration that instilling such a commitment in surrogate mothers could lessen the chance of their breaching contract terms that were included primarily to protect the fetus.

The issue of class and relative power is involved when a breach-of-contract issue is litigated. The most persuasive lawyers, often the most expensive, can not only win various contract disputes but can also use the very breach as evidence in the custody portion of a trial, as was the case in the Baby M trial. Much of the testimony by lawyers and paid mental health professionals who worked for the Sterns was designed to show that the contract terms should be enforced because Mrs. Whitehead knowingly and willingly participated in the arrangement. During the custody phase of the trial, however, similar experts painted a picture of an unstable, insincere, and chronically untruthful Mary Beth Whitehead.

Thus, contract terms might turn out to be only as good as the lawyers and other experts a person is able to employ to defend his/her position in court. Moreover, whether "wins" are more than merely symbolic may depend on what remedies courts might enforce and why. It is beyond the scope of this writing to list the numerous possibilities for breach of the contract, along with the various remedies a judge might consider. However, a case in point is the awarding of custody and visitation rights after a surrogate like Mrs. Whitehead refuses to relinquish the child. Her family's debts, her husband's occupation (sanitation engineer), and the family's yearly income ($28,000) were repeatedly brought up in the trial as a major handicap to her "fitness" to parent the child whose custody was being contested.

Postagreement: Gestation and Birth

I have combined the gestation and birth stages because they are inextricably connected as they relate to the issues of bonding, emotional attachment, and the relationship that develops between a growing fetus and its gestator. It is during these stages that, in spite of contract terms, actions by the surrogate mother can bring her personal desires and perception of her rights into direct conflict with the rights of the developing fetus. This, of course, presumes that the fetus has rights.

Mrs. Whitehead experienced few difficulties during her pregnancy; however, reports of her 50-pound weight gain, the phlebitis that she suffered from, and "other complications" suggest that future contracts may attempt to set limits on weight gain and other activities that are deemed by the contracting couple and its representatives as posing threats to the fetus. Also, the birth process might involve unanticipated emergencies that require fast and uncontracted-for actions by medical personnel (such as a forced cesarean section). Whether all such possibilities can be meaningfully included in the contract is highly unlikely. Therefore, two important questions are: Who may set limits and other behavioral constraints during pregnancy? And under what circumstances may conditions in the contract related to the birth process be imposed upon the surrogate mother against her will? An obvious answer to the first question is that the party with the most clever negotiator is likely to prevail in getting his or her terms into the contract. However, the issue of enforceability must be addressed in order to answer the second question.

Enforceability of contract terms during these stages is extremely tenuous. Short of the use of some type of imprisonment and behavior-control techniques that would undoubtedly violate the pregnant woman's fundamental rights to privacy and to the control of her own body, there is no acceptable way to safeguard and guarantee the respective rights of the pregnant woman and of the fetus that she is carrying. There is no way to ensure at the time the contract is signed that the surrogate will, during gestation and birth, feel an obligation to, or responsibility for, the fetus. A breach of contract terms during these two stages may be viewed as a justified assertion of a woman's fundamental rights, or as a criminal offense, such as battery against the fetus, or as murder.

Remedies for breach of contract terms by the surrogate mother during gestation and birth must be viewed as totally inadequate if the injury or death of the fetus resulted from the breach. Although criminalizing certain acts of the surrogate, levying a stiff fine upon her, or making her pay restitution to the couple may satisfy outside observers who espouse platitudes about social justice, such sanctions will not gain for the couple the benefit of their transactions. Such remedies cannot compensate the couple for the injury or destruction of the fetus. On the other hand, breach of contract by the couple, such as a refusal to pay the surrogate's fee, has far less potential for disastrous results and is decidedly more amiable to remediation through an order for specific performance or payment.

Psychological testimony during the Baby M trial suggests that society has not satisfactorily defined or come to terms with what mother-fetus or mother-child bonding really is. Moreover, the very term "mother" has been called into question now that it is possible for the biological, gestational, and social/adoptive mothers to be different women who all claim the same undefined rights to the child. Mrs. Whitehead's testimony to a local newspaper was, "I signed on an egg. I didn't sign on a baby girl, a clone of my other little girl. . . . In the delivery room as my baby girl was being born, I knew I was not going to give her up" (National Law Journal 1986:8).

Although some commentators view such testimony sympathetically, others—such as a testifying psychiatrist—have concluded that such comments, and the fact that Mrs. Whitehead styles her twelve-year-old daughter's hair every day, show that she is "overenmeshed" with her children. The fact that Mrs. Whitehead was allowed to nurse Baby M and to take her home for three days has raised questions relating to bonding and whether the relationships that develop during the gestation period, the birth process, and as a result of the nursing experience should have any bearing on the resolution of disputes in surrogate mothering transactions.

Relinquishment and Adoption

Surrogacy, at this stage of its development, does not accommodate the act whereby a surrogate changes her mind about terminating her parental rights, particularly if a binding contract exists. This stage, in which the surrogate terminates her parental rights, may be the most complex in terms of the psychological dynamics involved.

Under the contract between Mary Beth Whitehead and William Stern, Whitehead agreed "that in the best interest of the child, she will not form or attempt to form a parent-child relationship with any child . . . she may conceive . . . and shall freely surrender custody to William Stern, Natural Father, immediately upon birth of the child, and terminate all parental rights to said child pursuant to this agreement" (Surrogate Parenting Agreement 1985).

Notwithstanding extensive writing and research on the development of parent-child relationships from biological, psychological, and social perspectives, little has been written on how to avoid forming such relationships that could guide judges who must consider similar contract clauses. This term begs for an analysis regarding the understanding of the mother-child relationship on the part of the brokers and attorneys who draft such clauses and an examination of the motivation and intention of any surrogate mother who would agree to such a term. Similarly, the obligation to "freely" surrender custody seems to assume that once a contract is signed, the issue of whether or not a surrogate mother can actually carry out such an act is no longer subject to debate. The Baby M case clearly shows that such an assumption is not the case.

Problems inherent at this stage of termination of parental rights demonstrate that contract drafting cannot occur solely within a legal perspective if the best interests of all parties are to be seriously considered and if there is a real intent to avoid litigation. Attorneys should hire mental health professionals early in the negotiations to counsel parties to the contract about potential traumatic psychological and emotional reactions in contested cases. It is too late once litigation has begun.

Couples who allow surrogate mothers to contract to do things that they believe may be psychologically impossible, comforted only by the knowledge that they have a better than average chance to win an ensuing court battle, are hardly dealing in good faith. Furthermore, surrogate mothers who agree to contract terms that involve psychosocial unknowns—because of their lack of foresight or inability to obtain legal counsel—should not automatically be condemned to undergo an assault in court from legal and mental health professionals hired by the couple, such as occurred in the Baby M trial.

Perhaps mediation or other alternative forms of dispute resolution are more suitable than adversarial trial for these complex arrangements in order to try to solve problems of enforcement and to provide adequate remedies for a wronged party. As the number of litigated cases increases, it is clear that predicting how judges will rule becomes impossible.

Payment

The most controversial issue at this stage is whether the Sterns-Whitehead transaction involved the sale of Baby M or was merely a service provided by Mrs. Whitehead for which she would have been compensated with $10,000. Mrs. Whitehead did not accept the $10,000 fee, and she did not complain about the payment of her medical expenses. However, future cases may present problems with the monies involved, as opposition to the commercialization of surrogacy increases and laws against it are passed. A case in point is the clause that provided that if the baby was stillborn the fee would be reduced to $1,000. This stipulation strongly suggests that a live child was being paid for.

The timing of payments can be crucial because of the requirements of some states' adoption laws, which prohibit the making of financial deals for a child before its birth and do not allow the exchange of money between private parties for adoptions. Where the money is placed, such as in an escrow account, and who is authorized to make payments for or to the surrogate mother may also have to be specified in well-drafted contracts by competent and experienced lawyers. The parties unable to hire such experts will be at a decided disadvantage in this stage of medical expense and fee determination.

The Decision

On March 31, 1987, Judge Harvey Sorkow, presiding judge of the Superior Court of New Jersey, Chancery Division, Family Part, Bergen County, decided the case, In the Matter of Baby M, a pseudonym for an actual person. The issue in the case read:

> The primary issue to be determined by this litigation is what are the best interests of a child until now called "Baby M." All other concerns raised by counsel constitute commentary. That commentary includes the need to determine if a unique arrangement between a man and a woman, unmarried to each other, creates a contract. If so, is the contract enforceable; and if so, by what criteria means and manner. If not, what are the rights and duties of the parties with regard to custody, visitation and support. [*In the Matter of Baby* M 1987:1]

The holding in the case read:

> This court enters judgement in favor of plaintiff as follows:
>
> 1) The surrogate parenting agreement of February 6, 1985, will be specifically enforced.
>
> 2) The prior order of the court giving temporary custody to Mr. Stern is herewith made permanent. Prior orders of visitation are vacated.
>
> 3) The parental rights of the defendent Mary Beth Whitehead are terminated.
>
> 4) Mr. Stern is formally adjudged the father of Melissa Stern.
>
> 5) The New Jersey Department of Health, Bureau of Vital Statistics and its ancillary and/or subordinate state or county agencies are directed to amend all records of birth to reflect the paternity and name of the child to be Melissa Stern.
>
> 6) Restraining the defendants, Mary Beth Whitehead, Richard Whitehead, Joseph Messer and Catherine Messer, their relatives, friends, agents, servants, employees or any person acting for and/or on their behalf, from interfering with the parental and custodial rights of the plaintiff, his wife or their agents, servants, employees or any other persons acting for and/or on their behalf.
>
> 7) Reserved to the plaintiff as heretofore ordered is their unpleaded claims for money damages.
>
> 8) Counsel for plaintiffs will submit a certification of services pursuant to R4:42-9 in support of their application for counsel fees.
>
> 9) The court will enter judgement against the defendants on all prayers for relief in the first and second courts of their counterclaim.

> 10) The Guardian ad Litem shall file certification of services pursuant to R4:42-9 to support her application for fees. She shall also submit to the court the statements of fees from her experts for allocation by the court.
>
> 11) The $10,000.00 being held by the Clerk of the Superior Court shall be the property of Mary Beth Whitehead.
>
> 12) The Guardian ad Litem shall be discharged herewith except for the purposes of appeal.
>
> And finally, all should listen again to the plea of the infant as voiced so poignantly by several of the professional witnesses, statements with which this court agrees to such an extent that it will use its total authority if required to accomplish these ends:
>
> It was said Melissa needs an end to litigation, she needs to have her parentage fixed, she needs protection from anyone who would threaten her protection so as not to be manipulated and she needs a strong support system to protect her privacy.
>
> It was also said it is not in Melissa's best interest to have such public attention and scrutiny of her life. Common sense and common decency require that what Melissa must learn, she must learn in a very private manner.
>
> Melissa needs stability and peace, so that she can be nurtured in a loving environment free from and sheltered from the public eye.
>
> This court says that Melissa deserved nothing less—stability and peace. [*In the Matter of Baby* M 1987:119–121]

After rendering his decision, Judge Sorkow called the Sterns into his chambers, where he conducted adoption proceedings that made Mrs. Stern the legal mother of Melissa.

It is beyond the scope of this chapter to thoroughly analyze or evaluate Judge Sorkow's judgment and comments. However, I will briefly discuss the following areas in terms of their possible implications for future cases involving parties from divergent socioeconomic classes.

1. Some specific allegations and requests were made by the defendants, Mr. and Mrs. Whitehead, and by Mrs. Whitehead's parents, Joseph and Catherine Messer. The Whitehead and Messer claims are presented in relation to the seven-stage analysis because these stages readily correspond with the contract and custody phases of the bifurcated, or two-pronged, trial that was held. Specifically, all of the allegations made by the Whiteheads concerned the enforceability of the contract, and Judge Sorkow's ruling and comments on them may be analyzed within the context of four of the seven stages: prenegotiations, preliminary negotiations, contract formation, and payment.

The remaining three stages—gestation, birth, and relinquishment and adoption—involve issues that were primarily addressed in the custody phase of the trial, such as maternal attachment, the rights of the maternal grandparents, and the effects on Mrs. Whitehead's other children. These issues were examined by Judge Sorkow from two key perspectives: the parents' patraie power of the state, or "that power of the sovereign to watch out over the interests of those who are incapable of protecting themselves," and the family-law concept of the best interest of the child.

2. Judge Sorkow ruled against the Whiteheads on each allegation related to contract issues and drew harsh conclusions about the characters of Mrs. Whitehead and her parents, the Messers, in spite of testimony by "experts" on their behalf during the custody phase of the trial.

3. Class differences between the litigants were significant.

The following outlines the claims and requests made by the Whiteheads and the Messers.

Claim 1: That the contract should be held unenforceable because the Whiteheads had no attorney at the time they entered the contract.

Prenegotiation. Parties' mental capabilities are key considerations during this stage, as noted by Judge Sorkow when he dismissed this claim by pointing out that there were no mental disabilities and that "it is well settled that disparity of education or sophistication is not considered grounds for avoidance of a contract" (*In the Matter of Baby* M 1987:80).

Thus, the onus to acquire competent counsel before contracting with the Sterns was on the Whiteheads, and their unwillingness or inability to do so are matters that the judge did not address. However, his comment about disparity in levels of education or sophistication is important to note. Throughout his opinion, that disparity was used, along with disparities in family stability and financial ability, to tip the scales in favor of the Sterns. Moreover, the comment indicates the apparent middle-class bias that was an integral part of the decision and comment. For instance, Judge Sorkow's determination that Mrs. Whitehead had performed a service, for which she agreed to accept a $10,000 fee, not only abruptly—and probably prematurely—settled the sales versus service issue; it also raises questions of class related to the possible ways that women in need of money might ultimately serve women of means in the area of procreation.

Claim 2: That the contract terms were unconscionable in that they were manifestly unfair, including the $10,000 fee, and that the allocation of risks were patently inequitable.

Preliminary negotiations; payment. Judge Sorkow found no unconscionability, and he dismissed the charge that Dr. Stern took no risks. He found no overreaching, disproportionate bargaining power or unfairness, thus no unconscionability, and he observed that Dr. Stern's risk was whether the child was normal or abnormal. The fee was vaguely addressed in terms of imprecise values for services and the fact that Mrs. Whitehead had agreed to the $10,000 sum. An adept negotiator fashions a balance between benefits and burdens within a framework of equal bargaining power.

Claim 3: That the contract should be rescinded because three types of fraud had been perpetrated by the Sterns: (1) nondisclosure of Mrs. Stern's infertility; (2) nondisclosure of Mrs. Stern's multiple sclerosis; (3) nondisclosure of the contents of the psychological evaluation by the Infertility Center of New York psychologist.

Preliminary negotiations. Judge Sorkow defined four elements of fraud as: (1) a material misrepresentation of a fact; (2) known to be false; (3) upon which a party relied; and (4) to its damage. After a lengthy discourse on the concept of infertility and the lack of consensus for its definition, he found no fraud. Further, he dismissed the charge related to the psychological tests by using a principal-agency analysis, whereby he wrote that the Whiteheads failed to show that the Sterns were principals who exercised control over their alleged agent, the Infertility Center.

Nondisclosure, it seems, would probably be minimized under the intense scrutiny of legal, psychosocial, and medical experts at this stage. A key issue that remains is whether the parties have the ability to hire counsel during the critical stages preceding conception; who will competently determine what must, and what need not, be contract terms; and who will later capably fight for the contract's enforcement.

Claim 4: That the contract is one of adhesion, with all the bargaining power resting with the Sterns.

Preliminary negotiations. Judge Sorkow wrote: "By definition, a contract of adhesion is one in which one party has no chance but to accept or reject the other parties terms and there are no options by which the party may obtain the product or service. Here, neither party has a superior bargaining position. Each had what the other wanted. A price for the service each was to perform was struck and a bargain reached. One did not force the other. Neither had expertise that left the other at a disadvantage. Neither had disproportionate bargaining power" (*In the Matter of Baby* M 1987:76). Notwithstanding the judge's words, two points are highly disputable: whether there was a contract for service, and whether the Sterns' superior knowledge of the law and their use of competent counsel disproportionately enhanced their bargaining power.

Claim 5: That the contract is illusory: Only Mrs. Whitehead has an obligation, whereas Dr. Stern only benefits.

Preliminary negotiations. Judge Sorkow wrote, "The fact is the contract does provide that there is an obligation and responsibility, that there is a life long responsibility by Mr. Stern for the child's support and welfare. The contract is not illusory" (*In the Matter of Baby* M 1987:88–89).

These five allegations reveal that the Whiteheads believed that they were manipulated and deceived during the prenegotiation and preliminary negotiation stages. These claims may also be reviewed under the agreement/contract formation stage, for the negotiator not only ensures that fair dealing has occurred but that only terms reflecting a conscionable and acceptable agreement are included in the contract. Even this cursory look at the judge's comments reveals the importance of having effective counsel prior to the onset of performance, which in this case were the attempts at insemination. His remarks strongly indicate that ignorance, ineptness of counsel, and lack of sophistication on one side do not readily translate into unconscionability and bad faith dealing on the part of the opposing, better informed and more effectively represented, party. Yet, Judge Sorkow's failure to give any credence to the possible effects of discrepancies in parties' knowledge, finances, and power leaves much room for future conflict based on class and emphasizes the importance for similar cases of the views judges hold on class membership.

Two additional requests made by Mrs. Whitehead and her parents may also be examined in relation to the contract stage, even though these requests were considered during the custody phase of the trial, in which the sole focus was the best interest of the child.

Claim 6: That Mrs. Whitehead should have a time period after delivery to determine if she wants to surrender the child.

Claim 7: That Mrs. Whitehead's parents should be granted the right to visit Baby M.

The custody phase of the trial was a battle between "expert" witnesses, and it culminated in what Judge Sorkow labeled, "an extraordinary judicial remedy"—the termination of Mrs. Whitehead's parental right. Testimony from both sides provided the judge with "clear and convincing evidence" that Mrs. Whitehead is "manipulative," "exploitive," and a "fawning" user of the media; possessive of "narcissistic ends"; "impulsive," with "emotional over-investment" in the baby; and a "woman without empathy." Her parents were labeled "unworthy" of their request. Should such a woman have been screened out of the surrogate mothering program? Did her "expert" witnesses serve her well? Was the factor that determined that Mrs. Whitehead would lose and that the Sterns would win the content of their charcters or the levels of their respective socioeconomic class standings?

The Aftermath

The decision of the Baby M case was followed by discussion and action in several quarters. An appeal was filed; the Vatican issued a doctrinal statement entitled "Instruction on Respect for Human Life in Its Origin and on the Dignity of Procreation: Replies to Certain Questions of the Day";[4] a survey of lawyer opinions was conducted;[5] and state legislatures met to consider a rash of bills proposed to regulate or outlaw surrogate parenting.[6]

These activities, which occurred during an ongoing national debate, helped to identify the moral, psychological, and legal aspects of surrogacy. Although each of these aspects is important, it has been the evolution of laws governing the process in general upon which most attention has been focused. Following are the most important issues emerging from the debate.[7]

In an article published before the appeal, Harold J. Cassidy, the attorney for surrogate mother Mary Beth Whitehead, summarized his arguments (Cassidy 1987:19–20):

1. Before the "nebulous" term "best interests" can be used to justly determine custodial arrangements in contested surrogate parenting cases, the legal standard to be applied must be clearly defined.
2. "Natural" custody resides in the mother and when two natural mothers are both seeking to be the primary custodian of the child, the best interests should not be made an either/or proposition.
3. One parent should not—and cannot—be effectively eliminated from the child's life.
4. The court should do what is possible to provide the child with the best of what each parent offers.
5. The unique facts of the Baby M case involve constitutional and public policy considerations that should be considered in setting the standard and defining "best interests of the child."
6. The fundamental rights of the child were violated when the father attempted to pay the mother to give up the primary custody of the baby.
7. Using the surrogate—regardless of whether it was for payment—as a human incubator and requiring her to break the natural and full relationship with her child is a form of exploitation.
8. The contract seeks to effectuate abandonment of a baby, contrary to our policy against abandonment and improvident separations.
9. The state should not be involved in making determinations as to whether white-collar, formally educated, wealthier parents make better parents than those who are blue-collar, less-educated, and not wealthy.
10. The state's order to change custody from the surrogate mother to the father is impermissible state involvement in improper private conduct and promotes unconstitutional conduct.

11. State intervention that interrupted the relationship between Mrs. Whitehead and Baby M violated her due process rights because she was not notified. This gave the Sterns an advantage.
12. Dr. Stern committed fraud by failing to disclose the psychological report that indicated that Mrs. Whitehead might not be able to give up the baby after birth.

Gary N. Skoloff and Edward J. O'Donnel, attorneys for the Sterns, also summarized their arguments in a preappeal article:

1. The typical surrogate parenting agreement may be operative as a valid surrender of parental rights and consent for adoption.
2. The surrogate parenting arrangement is the creation of new life in accordance with society's respect and reverence for the family unit.
3. The right of a married couple to engage in collaborative reproductive techniques such as surrogate parenting is constitutionally protected.
4. It must be reasoned that if one has a right to procreate coitally, then one has the right to reproduce noncoitally. If it is the reproduction that is protected, then the means of reproduction are also to be protected.
5. A parent's right to the companionship of his or her child, when asserted, is potent. Nevertheless, it is recognized that an individual may knowingly and voluntarily waive a constitutional right.
6. The best interests of children involved in custodial disputes have been found to transcend all other considerations. When called on to specifically enforce a surrogacy contract, the court's determination that enforcement would not be inimical to the child's welfare is essential.
7. Given the right to procreate is fundamental, governments cannot unduly interfere with the procreative liberty of participants to a surrogacy arrangement. They are bound by the Constitution to specifically enforce surrogacy contracts except where a countervailing compelling state interest is apparent. [Skoloff and O'Donnell 1987:18]

The arguments described set the stage for another court battle over Baby M. This culminated on February 3, 1988, when the Supreme Court of New Jersey overruled the lower court decision. The New Jersey Supreme Court found the surrogacy contract unenforceable and returned to Mary Beth Whitehead unsupervised visitation rights. The Sterns maintained custody of the infant. According to the New Jersey Supreme Court, commercial surrogacy is baby selling, and no such contracts can be upheld in a court of law. The appeal of the decision is still pending.

The case may well be unresolved until it reaches the Supreme Court of the United States; even then the resolution will only be legal. The issues raised by this, and other surrogacy cases, cannot be well resolved once the child is born. If surrogacy contracts are to exist, they must be well crafted before conception occurs for the protection of each individual involved.

Such a requirement may seem to ask a great deal, but considering the possible consequences of surrogacy arrangements, it is the least that can be done.

Notes

1. In addition to the Baby M case, the surrogate mother of a baby girl born on June 25, 1986, fought to keep the child and was awarded joint custody in California (*Munoz v. Haro* 1987).

2. Two examples of interfamilial arrangements are South Africa's "surrogate granny," which involved a forty-eight-year-old woman who agreed to carry to term an embryo that resulted from the fertilization of her daughter's egg with the sperm of her son-in-law (Clairborne 1987:A38), and a forty-six-year-old newly married American woman who allowed her daughter to donate a egg for in vitro fertilization by the sperm of the mother's new husband (Andrews 1985:29–30). Although these arrangements have not yet resulted in complex legal battles, questions related to the rights and responsibilities of all parties and concerning how such arrangements will affect what we now call "the family" may be anticipated.

3. In situations in which parties to a contract bargain are unequal in their capacities, the process of the bargain and the substance of the bargain are in question. Courts often refuse to enforce, or they rescind, agreements in which one party has a highly unequal advantage, or in which there have been unfair or deceitful negotiations (Farnsworth and Young 1980:349–350).

4. In the Vatican statement, church officials condemned such increasingly common practices as artificial insemination, surrogate motherhood, and in vitro, or test-tube, fertilization—even when the sperm and ovum are supplied by husband and wife (Ratzinger and Bovone 1987).

5. In a poll conducted for the American Bar Association Journal, 601 lawyers were telephoned in March 1987. Lawyers agreed (60 percent, as compared to 25 percent who did not agree) that surrogate mothers such as Mary Beth Whitehead should have no legal claim to custody of children they are paid to bear: and 51 percent thought that there should be a thirty-day "grace period" after birth, during which a surrogate mother could change her mind about the agreement and decide to keep the child (Reidinger 1987).

6. After the Baby M decision, several state legislatures held hearings, and their members introduced a variety of bills that would either outlaw surrogacy altogether or regulate and redefine certain aspects of the practice (Andrews 1987).

7. *Tampe Tribune*, July 18, 1987, p. 4A.

References

Andrews, Lori. 1985. When Baby's Mother Is Also Grandma and Sister. Hastings Center Report 15, no. 5:29–30.

______. 1987. The Aftermath of Baby M: Proposed State Laws on Surrogate Motherhood. Hastings Center Report 17, no. 5:31–40.

Cassidy, H. J. 1987. What Is in Baby M's "Best Interest" and What Standard Should Apply? National Law Journal 10 (October 26):19–20.
Claiborne, William. 1987. "Surrogate Granny" Sparks Debate in South Africa. Washington Post, April 12, p. A38.
Farnsworth, E. A., and William F. Young. 1980. Contracts: Cases and Materials. Mineola: The Foundation Press, pp. 349–350.
In the Matter of Baby M. 1987. No. FM-25314-86 E.
Levine, Judith. 1986. Whose Baby Is It? Village Voice 31 (November 25):15.
Munoz v. Haro. 1987. Civ. 572–834.
National Law Journal. 1986. Vol. 9, no. 3 (September 29):8, col. 3.
Ratzinger, Joseph, and Alberto Bovone. 1987. Text of Vatican's doctrinal statement on human reproduction. New York Times, March 11, p. A14–21.
Reidinger, Paul. 1987. Lawyers Reject Surrogate Mother's Claim: Lawpoll. ABA Journal (June 1), p. 55.
Skoloff, G. N., and E. J. O'Donnell. 1987. Is Surrogate Parenting the "Cure" for Society's Infertility "Epidemic"? National Law Journal 10 (November 2):18.
Surrogate Parenting Contract. 1985. (February 6).

The Contributors

Renee R. Anspach
Assistant Professor
Department of Sociology
University of Michigan

Dick Batten
Doctoral Student
Boston College

Susan Borg
Freelance Architect

Sandra Anderson Garcia
Associate Professor
Department of Psychology
University of South Florida
Lawyer

Michael A. Grodin
Director Pediatric Walk-in Clinic
Boston City Hospital

Jeanne Harley Guillemin
Professor
Department of Sociology
Boston College

Susan L. Irwin
Doctoral Student
Department of Anthropology
Michigan State University

Brigitte Jordan
Professor
Department of Anthropology
Michigan State University

Judith N. Lasker
Associate Professor
Department of Sociology
Lehigh University

Betty Wolder Levin
Associate Professor
Department of Health and Nutrition Sciences
Brooklyn College
City University of New York

Kamran S. Moghissi
Professor
Department Obstetrics/Gynecology
Wayne State University

Marilyn L. Poland
Associate Professor
Department Obstetrics/Gynecology
Wayne State University

Rayna Rapp
Associate Professor of Anthropology
Graduate Faculty of Political and Social Sciences
New School for Social Research

I. David Todres
Director Pediatric Intensive Care
Massachusetts General Hospital

Linda M. Whiteford
Associate Professor
Medical Track Leader
Department of Anthropology
University of South Florida

Sybil Wolfram
University Lecturer in Philosophy
Fellow and Tutor
University of Oxford (UK)

Index

MONTGOMERY COLLEGE LIBRARIES
0 0000 00499229 3